Beyond Science and Empire

Through ten case studies by international specialists, this book investigates the circulation and production of scientific knowledge between 1750 and 1945 in the fields of agriculture, astronomy, botany, cartography, medicine, statistics, and zoology.

In this period, most of the world was under some form of imperial control, while science emerged as a discrete field of activity. What was the relationship between empire and science? Was science just an instrument for imperial domination? While such guiding questions place the book in the tradition of science and empire studies, it offers a fresh perspective in dialogue with global history and circulatory approaches. The book demonstrates, not by theoretical discourse but through detailed historical case studies, that the adoption of a global scale of analysis or an emphasis on circulatory processes does not entail analytical vagueness, diffusionism in disguise, or complacency with imperialism. The chapters show scientific knowledge emerging from the actions of little-known individuals moving across several Empires—European, Asian, and South American alike—in unanticipated places and institutions, and through complex processes of exchange, competition, collaboration, and circulation of knowledge.

The book will interest scholars and undergraduate and graduate students concerned with the connections between the history of science, imperial history, and global history.

Matheus Alves Duarte da Silva is a Postdoctoral Research Fellow at the University of St Andrews, working on the global history of medicine. He is the author of *Quand la peste connectait le monde: production et circulation de savoirs microbiologiques entre Brésil, Inde et France (1894–1922)* (2020).

Thomás A. S. Haddad is an Associate Professor of History of Science at the University of São Paulo, Brazil, specializing on astral knowledge practices in early modern empires. He is the author of *Maps of the Moon: Lunar Cartography from the Seventeenth Century to the Space Age* (2019).

Kapil Raj is a Distinguished Research Professor at the École des Hautes Études en Sciences Sociales, Paris, whose research is focused on the role of intercultural encounters in the construction of modern science. He is the author of *Relocating Modern Science: Circulation and the Construction of Knowledge in South Asia and Europe, 1650–1900* (2007).

Empire and the Making of the Modern World, 1650–2000

This monograph series seeks to explore the complexities of the relationships among empires, modernity and global history. In so doing, it wishes to challenge the orthodoxy that the experience of modernity was located exclusively in the west, and that the non-western world was brought into the modern age through conquest, mimicry and association. To the contrary, modernity had its origins in the interaction between the two worlds.

Portuguese Colonial Cities in the Early Modern World
Edited by Liam Matthew Brockey

Art in the Time of Colony
Khadija von Zinnenburg Carroll

Archiving Settler Colonialism
Culture, Space and Race
Edited by Yu-ting Huang and Rebecca Weaver-Hightower

Decolonising Europe?
Popular Responses to the End of Empire
Edited by Berny Sèbe and Matthew G. Stanard

Anti-Slavery and Australia
No Slavery in a Free Land?
Jane Lydon

Beyond Science and Empire
Circulation of Knowledge in an Age of Global Empires, 1750–1945
Edited by Matheus Alves Duarte da Silva, Thomás A. S. Haddad, and Kapil Raj

For more information about this series, please visit: https://www.routledge.com/
Empire-and-the-Making-of-the-Modern-World-1650-2000/book-series/EMPIREMOD

Beyond Science and Empire

Circulation of Knowledge in an Age of
Global Empires, 1750–1945

**Edited by Matheus Alves Duarte da Silva,
Thomás A. S. Haddad, and Kapil Raj**

Routledge
Taylor & Francis Group

LONDON AND NEW YORK

First published 2024
by Routledge
4 Park Square, Milton Park, Abingdon, Oxon OX14 4RN

and by Routledge
605 Third Avenue, New York, NY 10158

Routledge is an imprint of the Taylor & Francis Group, an informa business

British Library Cataloguing-in-Publication Data
A catalogue record for this book is available from the British Library

ISBN: 978-0-367-41072-8 (hbk)
ISBN: 978-1-032-55365-8 (pbk)
ISBN: 978-0-367-81454-0 (ebk)

DOI: 10.4324/9780367814540

Typeset in Times New Roman
by KnowledgeWorks Global Ltd.

Contents

Figures

Contributors

Mauro Capocci is a researcher at the University of Pisa and teaches history of science and medicine. He has worked on the history of contemporary biomedical sciences, including genetics, molecular biology and the introduction and circulation of antibiotics.

Daniele Cozzoli is an Associate Professor at Pompeu Fabra University, Barcelona. He has worked on the history of optics, the history of biomedicine in post-war Italy, and Italian historiography in the twentieth century. His publications include *Il metodo di Descartes,* 2008, and *Ho avuto solo molta fortuna: Biografia intellettuale di Daniel Bovet*, 2016.

Daniel Dutra Coelho Braga is currently a Postdoctoral Researcher at the Department of History at the University of São Paulo (USP – Brazil). In 2020, the Brazilian Society for the History of Science awarded him the SBHC Biannual Award for Best Thesis for his doctoral dissertation, defended in the Graduate Program in Social History of the Federal University of Rio de Janeiro in 2019. He was a visiting doctoral scholar at the Centre Alexandré Koyré (CAK-EHESS) in Paris, in 2017, and at the Rachel Carson Center for Environment and Society (RCC-LMU) in Munich, in 2018.

Jordan Goodman is an Honorary Research Associate in the Department of Science and Technology Studies, University College London. His latest book, *Planting the World: Joseph Banks and His Collectors – An Adventurous History of Botany*, was published in 2020 (William Collins/HarperCollins).

Elizabeth Haines works at the National Archives, UK. Previously she was Vice-Chancellor's Fellow in History at the University of Bristol. She received her PhD in Geography from Royal Holloway, University of London. Her research explores the practices of knowledge-making, particularly in colonial and post-colonial contexts.

Stacie A. Kent is an Assistant Professor at Boston College. Trained in history and social theory at the University of Chicago, she works on global histories of capital accumulation. Her work appears in Treaty Ports in Modern China (Routledge 2016) and Critical Historical Studies (Spring 2020). Her current book manuscript is "Coercive Commerce: Capitalism and Governance in the Late Qing, 1842–1911."

Lorelai Kury is a Research Professor at Casa de Oswaldo Cruz/Oswaldo Cruz Foundation, the Brazilian center for the history of science, medicine, and health, in Rio de Janeiro. She holds a PhD from École des Hautes Études en Sciences Sociale (EHESS), Paris, France. She has been a visiting scholar at King's College London, UK, at École des Hautes Études en Sciences Sociales, France, Universidade de Évora, Portugal, and at Universitat Autònoma de Barcelona, Spain. She has published several monographs, edited collections, articles, and book chapters on naturalist travelers along the eighteenth and nineteenth centuries.

Heloisa Meireles Gesteira is a Research Professor at the Museum of Astronomy and Related Sciences in Rio de Janeiro, Brazil. She has been a postdoctoral researcher at the History of Science Museum, University of Oxford, and has published articles and book chapters on the role of astronomy in Portuguese territorial claims in South America.

Shiori Nosaka is a PhD student at École des Hautes Études en Sciences Sociales (EHESS), Paris, France. Her dissertation investigates the history of bacteriology and epidemic control in Japan by contextualizing it within a framework of international technoscientific regimes in the late nineteenth century and early twentieth century.

Johanna Skurnik is an Academy of Finland postdoctoral researcher at the Department of Philosophy, History and Art Studies at the University of Helsinki. She received her PhD in European and World History from the University of Turku. She specializes in the history of geographical knowledge and science, map history and colonial history.

Alexander van Wickeren is a Postdoctoral Researcher at the a.r.t.e.s. Graduate School for the Humanities, University of Cologne. His book *Wissensräume im Wandel. Eine Geschichte der deutsch-französischen Tabakforschung (1780–1870)* was published in spring 2020 by Böhlau Verlag.

1 Science and Empire

Past and Present Questions

Matheus Alves Duarte da Silva,
Thomás A. S. Haddad, and Kapil Raj

The issue of May 5, 1967, of *Science* magazine, the weekly journal of the American Association for the Advancement of Science, carried an article by a young professor of history at the University of Texas at Austin named George Basalla. Titled "The Spread of Western Science," Basalla's paper tersely framed its assumptions and the problem it set to address already in the first paragraph: "A small circle of Western European nations provided the original home for modern science during the 16th and 17th centuries (…). Historians of science have often attempted to explain why modern science first emerged within the narrow boundaries of Western Europe, but few if any of them have considered the questions which is central to this article: How did modern science diffuse from Western Europe and find its place in the rest of the world?" The answer to this question, according to Basalla, was to be found in "a three-stage model [that] describes the introduction of modern science into any non-European nation."[1] In this model, European colonial and imperial expansion was the first step, because it paved the way for modern science to diffuse throughout the world. In the second stage, because of economic and strategical needs, or simply out of curiosity about the conquered territories, European empires created scientific institutions in their colonies, which became thereafter locales that acted as seeds of future scientific independence. The latter was the final step of the process, and even if it was to be achieved in parallel with political independence, it only reinforced the communion of former colonies and metropoles in the unitary values and pursuits of science. Indeed, Basalla's model framed "modern science" as a discrete entity with self-evident European origins, capable of attaining universal acceptance by virtue of its epistemic superiority and its inherently robust rules of institutional organization. Empires simply offered an infrastructure of sorts, or were the vectors, for the irresistible diffusion of such a highly successful way of producing knowledge and mastering nature and resources.

That Basalla published his paper in *Science*, a journal ostensibly featuring articles in the natural sciences, and not in a history of science venue, might come as a mild surprise, but the reason for it is as straightforward as it is telling. In the 1960s as today, *Science* had among its missions to influence policymaking.[2] Basalla's model echoed U.S. strategic goals of diffusing modern capitalism, as spelled out in Walter Rostow's model of development.[3] Indeed, Basalla's pitch to Washington, D.C.

DOI: 10.4324/9780367814540-1

policymakers was that training the scientific elites of what was at the time called the Third World and socializing them in the values of western (therefore "universal") science was an extremely important Cold War battlefield. Furthermore, Basalla's historically inspired blueprint for U.S. science-based hegemony was not at odds with the thinking that was going on in the "Third World" itself or in the communist bloc. In that era of revolutionary independence struggles, former European colonies were seemingly eager to acquire scientific expertise, skills, and institutions from past metropoles, and the Soviet Union was also keen on asserting its power through scientific knowledge. Science was regarded as a path to liberation, with no possible role in the violence and dispossession perpetrated by colonial rule, a thought that seemed extraneous to everybody's political agendas, including Basalla's. He was not interested even in arguing for a supposedly positive trade-off between colonial economic exploitation and political oppression and its "legacy" (scientific, technological, cultural), a point that would be made over the following decades and that occasionally resurfaces to the present day.[4] In short, for Basalla, empires were value-free drivers of the global spread of modern science.

From the early 1980s onward, Basalla's science and empire nexus became a crucial topic of contention among scholars. Reaction to "The Spread of Western Science" was indeed to become the focus of critique that significantly contributed, along with the continued interest in Joseph Needham's voluminous writings, to shape discourse on the "globalization" of science that emerged in the following decades.[5] The inextricable entanglement of the history of modern science with the history of imperial expansion and imperial social orders that organized the world over the last few centuries has by now arguably become common knowledge among historians of both science and empire.[6] In the introduction to a recently published handbook on science and empire, for example, the book's editor Andrew Goss refers several times to the "imperial turn in the history of science."[7] One can trace back the roots of this turn to works that started to be undertaken almost four decades ago, in no small measure as a reaction to the kind of Eurocentric (and mute on the effects of colonialism and imperialism, if not benevolent) assumptions that informed Basalla's model and much of the history of science at the time.

Indeed, in the canonical narratives that structured the discipline in the U.S. and Europe in the mid-twentieth century, such as the works of Alexandre Koyré, Rupert Hall, or Thomas Kuhn, northwestern Europe ("the West") is uniformly depicted the fountainhead of modern science.[8] Ignored are not only parallel processes carried out in "the Rest" but also the role of European colonial and imperial expansion in the emergence of modern science. The converse, however, is not true: Europe's supposedly superior scientific resources have been taken as one or the main explanation of the West's apparent rise after the fifteenth century. Even scholars highly critical of European colonialism have been willing to view science as the practical (but not moral) enabler of western claims to domination.[9] However, judging by the steady output of high-quality books, articles, and conferences devoted to the scrutiny of aspects of the relationship between science and empire over the last several decades – ranging from the most detailed micro-historical case studies to attempts at grand theorizing – it is patent how absolutely mainstream the theme has become in the history of science.[10] Even more significantly, in places such as Latin America, empires are no less than the core of scholarship in the field. Works produced by scholars in or of the region address topics that range from the role of the early modern Iberian empires in the construction of a framework for the diffusion or circulation of knowledge to the emergence of national scientific traditions in a context marked by neo-colonial relationships between independent Latin American countries and France, Britain, and, more recently, the U.S.[11]

Taking into consideration the previous observations about the centrality of empires to the modern history of science, we suggest that an examination of one of the numerous questions raised by its emergence, namely the methodological one, may be a fruitful way of engaging with the "imperial turn." Indeed, the chapters that the reader will find in this book are not unified by a single scientific theme, time period, or geographical region: the chapters cover approximately 200 years, and range from studies of British mapmaking in Africa and Oceania to late-Qing commercial statistics, from agricultural improvement in the Napoleonic Rhineland to zoological gardens in fascist/imperial Italy to a global, trans-imperial race to produce vaccines and sera against the bubonic plague. As discussed in detail below, their common thread lies in the explicit acknowledgment that these different case studies can serve as novel methodological propositions for investigating the science and empire nexus in light of contemporaneous discussions on global history. Therefore, by showcasing the problems, challenges, and potentialities inherent in the use of tools brought from global or relational studies, especially the idea of the circulation of knowledge, the chapters intend to refresh the already long and multifactorial trajectories of research on science and empire.

Trajectories of Science and Empires

The imperial turn, whatever it really means and whenever it actually took place, is not anywhere close to possessing a unified, uncontested meaning. The dyad "science and empire" is, in important ways, a kind of wildcard or floating signifier that can be attached to very different historiographical attitudes, theoretical or methodological practices, and research agendas. Perhaps most importantly, science and empire can be related to larger questions, such as: how were imperial orders and science imbricated with each other in the early modern and modern periods? Were empires and the sciences united in a process of co-construction, or did the one, even as it was influenced by the other, dominate the latter's development? Multiple and contradictory answers to these interrogations have been given by historians of science over the last 40 years, depending on the meanings they attributed to the very constitutive terms: "science," "empire," "modern," and so on.

To advance a historical – but not genealogical – account of the historiography of science and empire, a good starting point is the one we have already anticipated above: the rapid surge in criticism of Basalla's model starting in the 1980s. This criticism came especially from a quite heterogeneous and loosely structured group of historians of science that started to meet by the middle of the decade under the banner of "Science and Empire." In foundational meetings held in India, Australia, and France and in edited proceedings that were to become quite influential, the group stressed, for instance, that advanced knowledge practices existed in the "Rest" before the arrival of Europeans, who were responsible notably for suppressing – if not outrightly stealing and repackaging – "traditional," indigenous, or local forms of knowledge and ways of knowing. Casting, in general, a very negative eye on European imperialism, historians of science associated with the group forcefully made the case that modern science was not a "gift" from Europe to the Rest, but much more frequently an imposition. Scholars also paid a great deal of attention to the ideological power of modern science in justifying British imperialism and the French *mission civilisatrice* in Africa and Asia while stressing the operative role of scientific practices such as cartography, statistics, and medicine in the maintenance of European imperial rule over those regions.[12] Nevertheless, despite these new trends and the important questions they raised, science's ultimate Europeanness went largely, and paradoxically, unquestioned.

The latter issue of science as an ideological or practical "tool" of European imperialism was independently pursued at length by scholars unrelated to the Science and Empire group, coming not from the history of science but from disciplines such as anthropology or the history of international relations. To limit ourselves to just two influential examples, we could mention Daniel Headrick's suggestively titled *The Tools of Empire*, from 1981, or Bernard Cohn's *Colonialism and its Forms of Knowledge*, published in 1996. Cohn's study is especially worth mentioning for his stress on British "colonial knowledge" about India, in particular in the fields of linguistics, law, and ethnography, which are viewed as far more instrumental to the imposition of the imperial order than the more usual subjects of the history of science, such as physics, botany, or even medicine.[13] Although not always acknowledged, Cohn's book anticipated in important ways another "turn" that has been taking hold of the history of science as of late, namely a move toward a refashioning of the whole discipline as a "history of knowledge" in the broadest sense.[14]

Other scholars, also independently from the Science and Empire collective and focusing mainly on the British and French empires, stressed the reconfigurations of experience undergone by metropolitan scientific disciplines thanks to the "novelties" brought from the colonies in the form of new artifacts, animals, plants, and data in general. For them, a (possibly) critical view of imperialism was entirely compatible with the recognition that modern science developed significantly through the work of individuals or European metropolitan institutions relying largely on data produced by "imperial agents," such as colonial administrators, missionaries, sailors, and merchants. Inspired directly by Bruno Latour's "centres of calculation" schema, the focus of this important body of work was largely on the ways and means of the intellectual and material practices developed by those metropolitan practitioners to deal with the "inscriptions" that reached them from the distant recesses of the empire, or on the experiences of colonial agents "in the field."[15] Thus, a collateral effect of this approach was that the "Rest" ended up being (once again) exoticized or relegated to the role of a repository of raw information, scattered pieces of data that were to become true knowledge only after being processed in European institutions. Even the explicit recognition of the violence inherent in the gathering of such bits of information reaffirmed the centrality of Europe in the process and the lack of agency on the part of highly constrained colonial subjects,[16] doomed to be mere informants of the true, European knowledge-producers.[17]

In parallel with the various approaches we have mentioned, a distinct approach to the study of the relationship between science and empire in Latin America began to emerge in the 1980s. This was due to at least three reasons. First, historically associated with the Spanish and Portuguese empires, the region partook from the very beginning to be confronted on a massive scale with modern European expansion and colonization, and conversely, the first, after the U.S., to deal with practical and epistemological problems and potentialities that emerged in a "post-colonial" order. Second, after the early-nineteenth-century independences, successor empires appeared in Latin America as well: the short-lived Mexican empires (1821–1823, 1864–1867) and the more durable Brazilian one (1822–1889). Finally, the region engaged in a neo-colonial relationship with the U.S. in the late nineteenth century, and the U.S. was not only depicted metaphorically as a new empire but acted as a *de facto* imperial power.[18]

History of science has long been practiced in Latin America, at least since the 1930s, when it first caught the attention of scientists in search of a past – and a future – for their own disciplines, memorialists, and intellectuals invested in the elusive quest for national identities. The establishment of the history of science journal *Quipu*, in 1984, in Mexico, represented an important landmark, because it triggered a wealth of innovative studies focused on the cultural and social aspects of science in Latin America.[19] The journal's title, a reference to an Andean device used to gather data and make sophisticated calculations in the Inca Empire and its tributary states, was already a symbolically charged decision. It criticized the assumption of the region as a scientific blank slate before the conquest and argued in favor of the existence of scientific thought in what was to become Latin America.[20] Influenced by debates on centers and peripheries, and dependency theory, the generation of Latin American scholars organized around *Quipu* was mainly interested in understanding the complexities of what they viewed as the process of transfer of modern science from Europe to Latin America. Critically engaging with Basalla's diffusionist model, these scholars showed that far from being smooth and linear, as Basalla suggested, the "spread of modern science" to Latin America was at times violent and often marked by the need to adapt and negotiate with alternative intellectual cosmologies.[21] The *Quipu* generation argued that there was no such a thing as a sole path to achieving scientific independence and modernity. Instead, they showed how social groups seldom pictured in the region as intellectually progressive, such as the Catholic church, landowners, or colonial officials, were instrumental in the creation of the first scientific institutions in the Iberian American colonies.[22]

Lastly, another focus of intense historiographical activity worth mentioning in this overview of some of the approaches to the study of the relationship between sciences and empires is the reassessment that has been taking place since at least the early 2000s of the place, or rather the absence thereof, of Spain and Portugal in the mainstream narratives of the Scientific Revolution. Many scholars have thus argued that the early demands of oceanic navigation and long-distance government of pluricontinental empires pushed sixteenth-century Iberian knowledge makers to anticipate many of the developments that would later be exclusively associated with (northwestern European) modern science: the mathematization of space, widespread use of measuring instruments, standardization of data-collection procedures, and centralization of knowledge-processing institutions. According to this view, the establishment and reproduction of Spanish and Portuguese imperial orders would thus have created the need for, and be totally dependent on, artifacts of knowledge such as geographical and maritime maps, systematic compendia of local fauna and flora, efficient metallurgical processes, and wide-ranging materia medica – all on a planetary scale never seen before. As much as one can dispute this narrative of Iberian precedence (which explicitly aims to disrupt the equally disputable narrative of the Scientific Revolution), one must recognize that it squarely situates these Iberian natural knowledge-making practices in the context of empire.[23]

Despite the heterogeneity of the responses given in the last few decades to the question of what have been, after all, the relations between science and empire, the various strains of work we have mentioned above generally converged on assuming that modern science was first developed in Europe and then it naturally diffused to, or was forcibly imposed on, or willingly transferred to, or creatively adapted by, the rest of the world in the wake of imperial expansion – if not as a precondition for it. In other words, even when those works did not view modern science as a positive object that fortunately spread from the West to the Rest, most of them tacitly endorsed Basalla's central assumption that northwestern Europe (and later the U.S.) was the ultimate source and instigator of modern science.

Science and Empire in the Age of Global History

More recently, historians of science seeking a dialogue with the burgeoning field of global history[24] and other relational approaches, such as *histoire croisée*,[25] connected,[26] circulatory,[27] and transnational histories,[28] have offered a fresh look at the relationship between science and empire. Investigating the detailed dynamics of the circulation of knowledge within and across imperial formations, some scholars have also stressed the crucial role played by thus far invisible agents – "go-betweens" – in bringing together disparate cultural, scientific, and technical *milieux*. They have thus attempted to show that the emergence of modern science should be understood as a global interactive and transactional process rather than as a simple diffusionist one. According to this view, modern science emerged not from a static place or "network" of places, but from processes that were dynamic: it was constructed "on the move," in and through encounters, interactions, negotiations, and disputes between agents across the globe, European and non-European alike. These agents operated across a range of scales, dictated at the same time by their local conditions and the global currents that traversed their lives.[29] Through their interactions, they ignited, transformed, or halted processes that can be viewed as fundamentally involving some kind of "movement" (even if only metaphorically): trading, translating, passing on or withholding information, collaborating, and resisting. European imperial expansion was a powerful force behind the creation of those global fluxes and the contexts that made interactions between agents across multiple scales possible. Nonetheless, many other non-European empires also brought together willing or unwilling agents to a position where they had no alternative but to interact with each other and to broker the interconnection of different scales of historical agency. The most important point to be made is that the knowledge and ways of knowing that such agents carried into their exchanges were in no way shielded from becoming something else in the aftermath. Agents and institutions located in the "Rest" of the world had an active and much larger than previously appreciated role in shaping what has usually passed as "European" knowledge.[30]

Nonetheless, these new approaches are not free from criticism. As stressed by Fa-ti Fan and other scholars, the insistence on "dynamical" aspects such as the circulation of knowledge and intercultural interactions can render invisible the inherent violence of everyday life in imperial social orders and ignore, for example, the all too real destruction of indigenous lives and forms of knowledge.[31] Speaking more generally about transnational approaches, the historian of science Dominique Pestre has argued that the focus on international networks and global flows is leading the history of science to undervalue or simply ignore the power structures that conform and shape global interactions.[32] In another vein, and making use of bibliometric insights, Stefanie Gänger has recently demonstrated the pervasiveness of the word "circulation" in fields ranging from globalization studies to the history of science. She has argued that the term has been often used strategically simply to replace former concepts, such as diffusion, transfer, or spread, without really challenging them or adding any layer of novelty or critical thinking.[33]

The chapters collected in this book attempt to answer these and several other criticisms or suspicions that have been legitimately raised in connection with approaches to science and empire inspired by global history approaches or by the notion of circulation. They do not shy away from pointing out the methodological limits of these positions (which are not panaceas, at any rate), but also, if anything, they show that it is perfectly possible to avoid the pitfalls feared, among others, by Fan, Pestre, or Gänger. The chapters demonstrate, not by theoretical discourse, but through detailed historical case studies, that the adoption of a global scale of analysis or an emphasis on circulatory processes does not entail analytical vagueness, diffusionism in disguise, or complacency with imperialism. In fact, the essays trace their origin to a symposium called "Spaces of Circulation and Colonial/Imperial Landscapes: Criticisms and Challenges," which we organized in September 2018 within the International Conference of the European Society for the History of Science that was held in London. Our main goal was to learn from more than 20 scholars coming from different parts of the world about the multiple meanings of "circulation of knowledge," whether as a concept, a tool, an approach, an attitude, or an object. Furthermore, we wanted to discuss not just the potentialities but also the limits and impasses that they experienced when making use of circulation and other global history-inspired notions in concrete investigations of science in an imperial world.

Ten of the papers presented in the 2018 symposium appear now in this book, in significantly reworked versions that reflect the fruitful discussions held during the meeting. They cover a diverse range of scientific disciplines – astronomy, botany, geography and mapmaking, statistics, medicine and microbiology, and zoology – and span a period that ranges from the end of the eighteenth century to the aftermath of World War II. Even though the geographical coverage of the contributions encompasses well-known imperial formations which have also been the focus of much previous work on science and empire, such as the British and French empires, the book lays a claim to novelty as it also directs scholarly attention to regions that have not figured in previous studies as prominently as British India. Thus, the reader will find chapters focusing on the Napoleonic empire in western-central Europe, the British in the Zambezi, or the oceans as imperial spaces per se, populated by ships and claimed by specific institutions. Other essays in the book shed light on European and non-European imperial formations (both colonial and non-colonial) that have attracted much less attention of historians of science writing in English – for example fascist Italy, Meiji-era Japan, late-Qing China, or the Brazilian nineteenth-century empire. The chapters show that phenomena already identified in previous studies, such as the asymmetries of power in imperial formations, the nature of intercultural interactions, and the use of more traditional forms of knowledge by metropolitan science, were not exclusive to European empires but followed similar patterns in Asian and South American configurations as well. Despite the geographical breadth of the chapters, readers might still notice important absences, most notably the Ottoman and Russian empires. Even with such absences, we are confident that the cases studied in the book already contribute to showcase original points of view and severely underrepresented geographies in English-language studies.

More importantly than striving for an impossibly complete geographical coverage, the essays collected in this book offer innovative, albeit necessarily partial, answers to the historical question of the relationship between sciences and empires, even as they bring to the fore the promises and challenges of a global history of science. Thus, the chapters ask questions such as the connections between scientific practitioners based on metropoles and colonies, the global circulation and construction of technologies, or the emergence of new scientific disciplines through global processes such as wars and pandemics. Despite their heterogeneity, a crucial commonality needs to be stressed: each chapter deals with very specific, situated subjects. These can be human actors, a scientific instrument, a book, or an institution. In describing them and following the connections they forged, materialized, or symbolized, the authors sketch the co-construction of empires, sciences, and processes of globalization. Instead of positing either the global or the empire as frames or scales existing *a priori*, the chapters strive to reconstruct the inextricable interplay between imperial structures and knowledge structures by placing archives, places, and objects "under the microscope." By attempting to adopt a global imagination on the scale of their subjects, the chapters thus contribute to recent discussions about the possible theoretical and methodological interactions between global histories and micro-histories, and about the consequences of these interactions for the history of science.[34]

To help the readers navigate across these simultaneously global and micro-histories of science, the book is divided in two parts: the first revolving around the broad problem of knowledge production on imperial "landscapes," and the second dealing with knowledge production at imperial "crossroads." By choosing the "landscape" metaphor, we sought to avoid flat or binary concepts and to insist on the complex topography of power-knowledge relationships inside any given imperial formation, a complexity that can only be revealed through a close and critical reading of the sources. "Crossroads," on the other hand, alludes to knowledge production occurring in places where different empires, or even empires and nation-states, interacted. These interactions happened because of the circulation of objects and humans or by the emergence of global conjunctures that connected different actors. Beyond simple metaphors, both terms also evoke concrete realities. The circulation of knowledge, as we have argued elsewhere, is not a fluid, unbounded phenomenon; it is defined by limits which are a manifold function of sociabilities, social, political, economic and imperial spaces, as well as by geography itself.[35] Thus, some authors examine the emergence of knowledge about spaces and territories by means of maps and health statistics, and how these spaces themselves conformed or shaped the scientific investigations carried on about or within them. In addition, some chapters examine how science was mobilized to create and justify political, economic, cultural, and ecological borders, while others pay attention to the uniqueness of scientific investigations carried on frontline spaces, such as customs, where cultural, linguistic, and social interactions and brokerage permanently took place.

Knowledge Production on Imperial Landscapes

The six chapters in this part focus mostly on the negotiation of power and scientific hierarchies within imperial formations and criticize a common understanding of the hegemony of European "centers" of knowledge. Widely promoted by Bruno Latour's idea of centers of calculation,[36] the assumption that people and institutions based on imperial metropoles exerted a kind of centripetal attraction, gathering information collected in other places and transforming it into stabilized, validated, and legitimate knowledge, found vast support ever since its formulation in the late 1980s.[37] Indeed, Latour's model offered a convincing solution to one of the main worries of the post-Kuhnian, more or less relativist strands of history of science that dominated the 1980s: how did modern science, which work after work demonstrated that was extremely localized in its practices, manage to attain universality? Latour's answer was that, on the one hand, centers of calculation enforced the standardization of data-collection processes that were initially recognized or deemed legitimate only locally. On the other hand, the knowledge the same centers created from the stable bits of information ("inscriptions") that reached them back-tracked in the form of standards, or immutable mobiles, to the rest of the world. In this back-and-forth, rigidly controlled by the centers, more and more people, institutions, and things were "enlisted" into an ever-growing "network" that took care of propagating – in an unmistakably diffusive manner – to the whole world what might once have been a legitimate way of knowing to just a handful of individuals in a metropolitan museum or laboratory. Highly malleable and free from any overt political commitment, Latour's model could easily be made to fit the interests and preoccupations of historians working on the relations between science and empire: centers of calculation equated metropolitan imperial institutions, "universalization" became another name for the process by which *European* science spread, traveled, or imposed itself throughout the world, science networks just followed imperial infrastructures (trade, administration, war), and so on.

Nonetheless, this approach has not been free from criticism, revision, or, to the very least, qualifications, a pattern explored by our authors. Thus, without denying the power asymmetries and the preeminent role of the historical actors based in European imperial institutions on the definition of the fates of uncountably many "colonial subjects," the chapters examine moments in which such an apparent uncontested hegemony was not at all evident, and when negotiations and unexpected outcomes were made possible. Indeed, the chapters in this part stress three main points. First, the limits of the power effectively held by places of knowledge based in Europe to impose their agendas on their colonies. Second, the emergence of colonial centers of knowledge formation and the role of agents based in the colonies in the production of knowledge that they strategically disclosed to or withheld from metropolitan counterparts to advance their own positions. Finally, as a complement to the latter point, the ability of actors in the colonies to reconfigure metropolitan endeavors for the sake of their personal and local projects.

Thus, Jordan Goodman's chapter discusses Joseph Banks's project to acclimatize breadfruit and other plants in different parts of the British empire. In following the journeys of these plants around the world at the end of the eighteenth century, Goodman invites us to a global travel that starts in Tahiti, approaches Jamaica and London, and ends in Calcutta. Goodman argues that the ships transporting the plants became not only "floating gardens" but places to produce new botanical knowledge. In this specific and unusual place, where Banks's directives were unenforceable, it was possible to measure the effects of climate and levels of humidity in the development of the plants, which enabled a better understanding of the feasibility and limits of their acclimatization. In showing this entangled process of material circulation and knowledge production, Goodman's chapter points out the role of actors commonly ignored in the history of botany, such as the captains and gardeners of the ships, and shows how the ships themselves were floating loci not only of British imperial social order but also of science.

In sequence, Alexander van Wickeren dismantles, through a carefully studied case, the widespread view that Parisian *savants* played a dominating and irresistibly centralizing role in planning the exploitation of natural resources in the Napoleonic empire. The chapter focuses on the cultivation of tobacco in the occupied Rhineland. The introduction of the state monopoly on tobacco from 1810 onward supposedly had the effect of empowering a Parisian-based scientific-administrative bureaucracy, whose best interests would be served by strict control of the production and circulation of tobacco and by the imposition of centrally devised agricultural improvements on the cultivators. However, van Wickeren demonstrates that cultivation lands were connected by entrenched regional regimes of agricultural knowledge circulation that predated the imperial annexation and became even more robust after it occurred. Parisian officials were wary of agricultural improvements for fear of generating surplus production and, as a result, contraband. In consequence, they unwittingly stimulated strategic alliances between provincial administrators and local agricultural societies that had been experimenting with innovative cultivation techniques for a long time. Thus, the chapter demonstrates how pockets of local agricultural knowledge thrived and developed in the face of an imperial order whose power to create something like a unified "French science" has thus far gone largely unquestioned.

Daniel Dutra Coelho Braga's chapter brings us back to maritime spaces through an investigation of the construction of medical categories of disease of the torrid zone and warm climates during missions of circumnavigation during the Bourbon Restoration (1815–1830) in France. By following the works of Pierre-François Kéraudren, head of the French Navy's Health Department, and his controversies and negotiations with French captains based in South American naval stations, Dutra shows how communication played a key role in the construction of French naval medical knowledge. Furthermore, Dutra argues that Kéraudren was able to address the challenges of overseas management, long-distance interlinkages, and local diseases not by claiming anything like the centrality that one might expect from Latour's model, but rather by leveraging on a low-key, middle-management position to become a kind of inextricable, one-man communication knot.

The next two chapters, by Johanna Skurnik and Elizabeth Haines, respectively, each deal with geographical knowledge and mapmaking in different parts of the British empire in the latter half of the nineteenth and early twentieth centuries. Skurnik examines in detail the ways in which the geographical knowledge deriving from Augustus C. Gregory's British government-funded North Australian Expedition (1855–1856) circulated, contrasting it with the modes of mobility of the material originating from John McDouall Stuart's privately funded expeditions between 1859 and 1863. Based on a wealth of sources that embody different forms of representation of information, such as letters, manuscript maps, printed maps, newspaper cuttings, and reports, Skurnik shows how the various media, as taken up in different places by different actors, affected claims of knowledge-making from the expeditions' materials. The author illustrates the social and material processes that enabled or hindered the circulation of knowledge between different spaces of British imperial knowledge-making about the geography of Australia. The central roles of little-known imperial civil servants in the metropolis and the colonies as important knowledge brokers who facilitated, controlled, and coordinated the movements of material containing geographical and cartographic information are revealed in the chapter.

Haine's chapter focuses on the careers of two British imperial agents working in the Zambezi valley (present-day Zambia) from approximately 1880 to 1920. Scholarship on the role of cartography in imperial projects, she argues, has tended to emphasize the importance of centralized knowledge for establishing and maintaining imperial power. Most studies thus concentrate on the metropolitan institutions that claimed ultimate authority over map-based geographical knowledge: institutions that ordered and tried to standardize the production of maps, custodied and controlled the circulation of an ever-growing amount of cartographic material according to metropolitan logics, interacted with other levels of government and the military, and established a symbiosis with geography as a scientific discipline. Yet, Haines demonstrates that as seen from the perspective of the field (rather than the metropole), the relationship between the circulation of knowledge and the exercise of authority looks very different. She makes the case that the journeys of the two protagonists of the chapter, although ostensibly to carry out mapping exercises of various kinds, are better read as political technologies in which the agents mediated between imperial intentions and local contexts. Their physical presence in the field was key to the enactment of both diplomatic and violent engagement with African political structures. More remarkably, Haines shows how both men exploited their local resources to act against the wishes of their superiors based in London, by controlling the level and quality of the information they were willing to disclose, in a striking example of an imperial landscape of power-knowledge in which the "center" is not where one would have usually expected it to be located.

Concluding the first part of the book, Mauro Capocci and Daniele Cozzoli's chapter follows the first decades of Rome's zoological garden (1909–1945). The authors analyze the early history of the institution both as part of a global trend of displaying and studying "exotic" animals found in imperial and national metropoles in several places, as well as its specific connections with the workings of Italian imperialism in Africa under Mussolini's Fascist regime. Capocci and Cozzoli criticize explanations for the grip of fascism that are based on Hannah Arendt's thesis about the nature of totalitarianism and argue that the regime needed to continuously justify itself before Italian society at the risk of losing mass support. The zoo thus became a showcase for the success of Italian imperial expansion by displaying animals brought from the Italian colonies in Africa and letting the Roman public know about their empire to support the enterprise. Contrary to the regime's expectations, though, the authors show that the Roman public was not utterly interested in the empire or at least in the animals, hence causing the zoo to sink into financial turmoil.

Knowledge Production at Imperial Crossroads

While the focus of the first part is on the construction of asymmetries, negotiated agencies, local powers, and circulation of knowledge within well-defined imperial formations, the four chapters in the second part examine intersections between different empires. The actors and places where those intersections occurred are numerous, and the authors do not aim to be exhaustive. Their main goal is to highlight what happened to the construction of knowledge when actors – humans or non-humans – crossed physical, political, and scientific borders, and, because of their movements, connected different locales. In that sense, the imperial crossroads examined in these chapters range from physical places that can be pinpointed on a map to metaphorical imperial frontiers that were either blurred or highlighted by the circulation of objects and the emergence of global conjunctures.

The second part begins with Heloisa Gesteira's examination of the presence of scientific instruments of British origin in Rio de Janeiro, Brazil, which emerges as key to understanding questions of competition and collaboration between astronomers working for different empires. Her narrative unfolds in August 1787, when a British Royal Navy fleet dropped anchor in Rio de Janeiro. On board of one of the ships was William Dawes, a British astronomer, colonial administrator, engineer, and topographer en route to establish an observatory in New South Wales, now Australia. During his very short stay, Dawes wrote to the British Astronomer Royal, Nevil Maskelyne, describing how he had made the acquaintance of Bento Sanches Dorta, a Portuguese astronomer who had lived in the capital of the Portuguese colony since 1781. Dorta had been sent there to help settle border disputes between Portugal and Spain around the Río de la Plata area. Gesteira takes Dawes's assessment of Dorta as the starting point for an investigation of how science was practiced in the key colonial capital of the Portuguese empire in the late eighteenth century, and how an agent of a foreign, at times rival power judged the work of a fellow astronomer and his instrumental skills.

Lorelai Kury's chapter takes us through the first decades following Brazil's independence from Portugal. The author uses the career of French botanist Auguste de Saint-Hilaire (1779–1853) as a window to the roles played by science in the ideological construction of Brazil's fledgling empire, founded in 1822, as well as in brokering the empire's cultural relationships with European powers, particularly France. Saint-Hilaire traveled across Brazil from 1816 to 1821, and all that he published in the years immediately following his return to Europe drew from the material he had previously collected. Kury shows how, at that moment, fierce competition to publish on Brazilian flora was underway in Europe, involving dozens of naturalists and collectors who had traveled throughout Brazil after 1815. Mastering all the available information pertaining to Brazilian natural history was increasingly challenging for European naturalists, and access to local experts was an invaluable resource. Through the careful reading of scores of Saint-Hilaire's publications and manuscripts, Kury demonstrates how his painstaking diligence in quoting and referencing Brazilian practitioners, Indigenous peoples, and laymen was part and parcel of the scientific and publishing effort. By so doing, Saint-Hilaire could claim he still enjoyed that coveted direct connection to local naturalists, enhancing then his scientific prestige in Europe as well as his relations with Brazil's new imperial elite. This elite in turn was also eager to find legitimation in post-Napoleonic France, becoming avid consumers of French cultural products – including, as Kury shows, Saint-Hilaire's botanical and travel reports, which found a strong market in Brazil in the 1830s.

Moving from Brazil to China, Stacie Kent presents a striking example of a powerful empire refusing to adopt a purportedly universal knowledge practice such as statistics. Her specific setting is a crossroads of nineteenth-century empires: the Imperial Maritime Customs, established in 1854 by the Qing dynasty to regulate trade relations between the Chinese empire, European powers, and the U.S. Although officially part of the Chinese government, the agency was headed by and staffed at the highest levels by Europeans and Americans. These foreign officials were particularly invested in the routine production, publication, and sale of exhaustive statistical reports on the volume and returns of trade, tabulating and enumerating aggregate values of commodities and ships passing through the Custom's offices. The striking fact is that while these statistical reports were a prized source of information outside China, particularly for the British Parliament, they were virtually ignored by Qing administrators for many decades, despite the reports supposedly obvious usefulness for fiscal matters, and in a moment when the Qing were interested in expanding state power through new governmental technologies. Through a careful reading of the sources and a sophisticated theoretical framework, Kent demonstrates that in the heart of the matter lay a fundamental disconnect between Qing officials' imperial practices and the world constituted by western statistics: for the Qing, the aggregate numbers generated by the Custom's statisticians simply did not correspond to, or failed to capture in any meaningful way, a real social order constituted by multiple discrete groups of producers and merchants, each one accountable to the government in widely different ways.

Linking Asia to South America, Shiori Nosaka and Matheus Duarte da Silva's chapter proposes a novel approach to frame the emergence and subsequent globalization of microbiology at the turn of the twentieth century. Rather than focusing solely on the diffusion of microbiological paradigms and techniques from Europe to the rest of the world, or on the establishment and institutionalization of the new science of microbes in European colonies, or even on the development of national microbiological traditions, they argue for a connected perspective. The chapter traces the first decades of the Third Plague Pandemic (1894–1950), a conjuncture that traversed different empires and independent nations. The authors map the global production and circulation of anti-plague sera and vaccines between three geographical units: the Japanese Empire, British India, and Brazil (by then a former empire, recently transformed into a republic). Global empires, especially the British, they contend, acted as spaces of circulation, where actors positioned outside these empires could strategically act by using the imperial infrastructures to pursue their projects. The authors show that the plague pandemic triggered several related processes of global reach, such as the establishment of new centers of production of microbiological knowledge around the world, the acceleration of the interconnections between these centers, and the development of new objects and techniques to treat and to immunize against plague. These developments, which took place mostly outside Europe, were central to the reconfiguration of the core of microbiology and impacted research carried on in European centers of the new science, particularly at the Pasteur Institute of Paris. Finally, the authors stress the capacity of the new centers to collaborate, but also to contest, redirect, or ignore the scientific agendas of European laboratories.

***

To conclude, here are a few remarks to guide our readers. It is worth noting that the authors do not adhere or propose any common or ultimate definition of what should constitute a global history of science nor of how to frame science and empire in the debates of global history. Rather, the chapters are empirical studies that display tools, questions, and methodologies that can inspire and provoke readers and ultimately produce fruitful engagements in the field. The answers offered throughout this book are thus essentially explorations, leaving the reader to judge for themselves the limits and potentialities of the strategies deployed by each author in dealing with the specific subject in their case study. Nonetheless, although each chapter can be read individually, reading them as a coherent ensemble will hopefully help better understand the critical engagement of our book with the "imperial turn" in the history of science.

Notes

1 George Basalla, "The Spread of Western Science," *Science* 156, no. 3775 (5 May 1967): 611–622, 611.
2 The same issue that carried Basalla's paper brought several shorter pieces with explicit policy content: one of them celebrated U.S. President Lyndon Johnson's reassuring remarks to the scientific community about the value of basic research, another discussed the new goals of French science policy, and another one reported on the views of the President's top scientific advisor on the importance of proper public understanding of science for the future of U.S. global leadership.
3 Walter W. Rostow, *The Stages of Economic Growth: A Non-Communist Manifesto* (Cambridge: Cambridge University Press, 1960).
4 For a sophisticated defense of the "civilizing mission" view of European colonialism, which is at the same time critical of its violent means, see Lewis Pyenson's trilogy: *Cultural Imperialism and Exact Sciences: German Expansion Overseas, 1900-1930* (New York, NY: Peter Lang, 1985); *Empire of Reason: Exact Sciences in Indonesia, 1840-1940* (Leiden: E. J. Brill, 1989); and *Civilizing Mission: Exact Sciences and French Overseas Expansion, 1830-1940* (Baltimore, MD: Johns Hopkins University Press, 1991). Cf. also the critique by Paolo Palladino and Michael Worboys, "Science and Imperialism," *Isis* 84, no. 1 (1993): 91–102, and Pyenson's rejoinder, "Cultural Imperialism and Exact Sciences Revisited," *Isis* 84, no. 1 (1993): 103–108.
5 Kapil Raj, "Beyond Postcolonialism...and Postpositivism: Circulation and the Global History of Science," *Isis* 104 (2013): 337–347. See also, Idem, "History of Science, Intellectual History, and the World, 1900-2020," in Stefanos Geroulanos and Gisèle Sapiro, eds., *Handbook of Intellectual History and Sociology* (New York, NY: Routledge, 2023), forthcoming.
6 To limit ourselves to just a few conspicuous examples, since 1997, the Division of History of Science and Technology of the International Union of History of Science and Technology actively maintains a "Science and Empire Commission" that holds conferences every four years within the International Congress of History of Science and Technology. Note also that two of the most widely read journals in the field, *Isis* and *Osiris*, have published dedicated issues on the theme quite some time ago: see Roy MacLeod, ed., *Nature and Empire: Science and the Colonial Enterprise*, in *Osiris*, 2nd series, 15 (2000); Sujit Sivasundaram, ed., *Global Histories of Science*, focus section of *Isis* 101, no. 1 (2010); and Londa Schiebinger, ed., *Colonial Science*, focus section of *Isis* 96, no. 1 (2015). Earlier still, there appeared edited collections that have been influential to the present day: Deepak Kumar, ed., *Science and Empire: Essays in Indian Context* (Delhi: Anamika Prakashan, 1991); Patrick Petitjean, Catherine Jami, and Anne Marie Moulin, eds., *Science and Empires: Historical Studies about Scientific Development and European Expansion* (Dordrecht: Kluwer Academic Publishers, 1992).
7 Andrew Goss, "Introduction: An Imperial Turn in the History of Science," in *The Routledge Handbook of Science and Empire*, ed. Andrew Goss (London: Routledge, 2021), 1–9. Quite surprisingly, however, in the essay that opens the book, Pratik Chakrabarti claims that "empire remains peripheral to history of science," an assertion that is to our

view unsupported by overwhelming evidence to the contrary. See Chakrabarti, "Situating the Empire in History of Science," in ibid., 10.

8 Alexandre Koyré, *From the Closed World to the Infinite Universe* (Baltimore, MD: Johns Hopkins University Press, 1957); Thomas Kuhn, *The Structure of Scientific Revolutions* (Chicago, IL and London: University of Chicago Press, 1962); A. Rupert Hall, *The Scientific Revolution, 1500-1800: The Formation of the Modern Scientific Attitude* (London and New York, NY: Longmans, Green, 1954).

9 Daniel R. Headrick, *The Tools of Empire: Technology and European Imperialism in the Nineteenth Century* (New York, NY: Oxford University Press, 1981); David Arnold, *Colonizing the Body: State Medicine and Epidemic Disease in Nineteenth-Century India* (Berkeley, CA: University of California Press, 1993). As will become clear, we are critical of this view of science as "tool" of domination, above all because it is still ultimately Eurocentric, even if in a negative way.

10 Obviously, this does not mean that the whole discipline of history of science has become fixated on science and empire, far from it. Even important works that set out to subvert the traditional, triumphant narrative of the Scientific Revolution may be entirely oblivious to whatever took place out of northwestern Europe, such as is the case with Steven Shapin, *The Scientific Revolution* (Chicago, IL: University of Chicago Press, 1996). Furthermore, it is also perfectly legitimate that a work concentrates on some European scientific development without being Eurocentric, for thematic or methodological reasons. What matters is that empire is in no way removed from the interests of many historians of science.

11 For recent reviews on Latin American historiography of science in global perspective, see Ana Barahona and Kapil Raj, "A Historiography of the Life Sciences and Medicine in Latin America in Global Perspective," and Matheus Alves Duarte da Silva and Marcos Cueto, "From the Social to the Global Turn in Latin American History of Science," in *Handbook of the Historiography of Latin American Studies on the Life Sciences and Medicine*, ed. Ana Barahona (Cham: Springer International Publishing, 2022), 1–15 and 19–38, respectively.

12 Cf. the various contributions in Kumar, ed., *Op. cit*; and Petitjean et al., eds., *Op. cit*.

13 Bernard S. Cohn, *Colonialism and its Forms of Knowledge: The British in India* (Princeton, NJ: Princeton University Press, 1996).

14 See, among many others, Jürgen Renn, "From the History of Science to the History of Knowledge – and Back," *Centaurus* 57, no. 1 (2015): 37–53; Peter Burke, *What is the History of Knowledge?* (Cambridge: Polity Press, 2016); Christian Joas, Fabian Krämer, and Kärin Nickelsen, eds., *History of Science or History of Knowledge?*, special issue *Berichte zur Wissenschaftsgeschichte* 42, nos. 2–3 (2019); Johan Östling, David Larsson Heidenblad, and Anna Nilsson Hammar, eds., *Forms of Knowledge: Developing the History of Knowledge* (Lund: Nordic Academic Press, 2020); and the *Journal for the History of Knowledge*, founded in 2020 by Gewina, the Belgian-Dutch Society for History of Science and Universities.

15 For centers of calculation, see Bruno Latour, *Science in Action: How to Follow Scientists and Engineers through Society* (Cambridge, MA: Harvard University Press, 2003), chap. 6.

16 A few representative examples of this approach are David Philip Miller and Peter Hanns Reill, eds., *Visions of Empire: Voyages, Botany, and Representations of Nature* (Cambridge: Cambridge University Press, 1996); John Gascoigne, *Science in the Service of the Empire: Joseph Banks, the British State and the Uses of Science in the Age of Revolution* (Cambridge: Cambridge University Press, 1998); Richard Drayton, *Nature's Government: Science, Imperial Britain, and the "Improvement" of the World* (New Haven, CT: Yale University Press, 2000); Londa Schiebinger and Claudia Swan, eds., *Colonial Botany: Science, Commerce, and Politics in the Early Modern World* (Philadelphia, PA: University of Pennsylvania Press, 2007).

17 A hierarchical, almost rigid distinction between "mere information" and "true knowledge" gained currency from works far removed from the history of science, most

notably Manuel Castells, *The Informational City: Information Technology, Economic Restructuring, and the Urban-Regional Process* (Oxford: Basil Blackwell, 1989).

18 Cf. Alfred W. McCoy, Francisco A. Scarano, eds., *Colonial Crucible: Empire in the Making of the Modern American State* (Madison, WI: University of Wisconsin Press, 2009).

19 Márcia Regina Barros Silva, "História e historiografia das ciências latino-americanas: *Quipu* (1984-2000)," *Revista Brasileira de História da Ciência* 7, no. 1 (2014): 47–57.

20 Juan José Saldaña, "Presentación," *Quipu: Revista Latinoamericana de Historia de las Ciencias y la Tecnología* 1, no. 1 (1984): 5–6.

21 It must be stressed that Basalla's model was criticized above all for its Cold War political underpinnings and for its lack of historical nuance in the representation of the process of diffusion of modern science from Europe to the Rest. However, the view of Europe as the (scientific) core and Latin America as a periphery remained largely unquestioned, being taken as a historical given and as a horizon for political struggle. As a result, the *Quipu* generation developed an undeniably rich conceptual vocabulary of creolization and appropriation of scientific knowledge and provided a trove of case studies of the creative local reception, by Latin American thinkers, of crucially important strains of modern science, such as Newtonianism or Darwinism – but Europe remained the ultimate source and arbiter. Examples of this approach that emphasized the complexities of reception are Luis Carlos Arboleda, "Acerca del problema de la difusión científica en la periferia: el caso de la física newtoniana en la Nueva Granada (1740-1820)," *Ideas y Valores* 38, no. 79 (1989): 3–26; and Heloisa Maria Bertol Domingues, Magali Romero Sá, and Thomas Glick, eds., *A Recepção do Darwinismo no Brasil* (Rio de Janeiro: Editora FIOCRUZ, 2003).

22 Edited collections that offer a good overview of this body of work are those of Antonio Lafuente, Alberto Elena, and María Luisa Ortega, eds., *Mundialización de la ciencia y cultura nacional* (Aranjuez: Doce Calles, 1993); Maria Amélia Dantes, ed., *Espaços da ciência no Brasil: 1800-1930* (Rio de Janeiro: Editora FIOCRUZ, 2001); and Juan José Saldaña, ed., *Science in Latin America: A History*, transl. Bernabé Madrigal (Austin, TX: University of Texas Press, 2006). For a critical reading of the achievements and limits of this generation of Latin American scholars, see Marcos Cueto and Matheus Alves Duarte da Silva, "Trayectorias y desafíos en la historiografía de la ciencia y de la medicina en América Latina," *Asclepio* 72, no. 2 (2020): 320.

23 A few collective works representative of this approach are Victor Navarro Brotóns and William Eamon, eds., *Más allá de la Leyenda Negra: España y la Revolución Científica* (Valencia: Universitat de València, 2007); Daniela Bleichmar, Paula De Vos, Kristine Huffine, and Kevin Sheehan, eds., *Science in the Spanish and Portuguese Empires, 1500-1800* (Stanford, CA: Stanford University Press, 2009); María M. Portuondo, ed., *Iberian Science: Reflections and Studies*, special issue of *History of Science* 55, no. 2 (2017); and Antonio Sánchez and Henrique Leitão, *Revisiting Early Modern Iberian Science, from the Fifteenth to the Seventeenth Centuries*, special issue of *Early Science and Medicine* 21, nos. 2–3 (2016).

24 Sebastian Conrad, *What Is Global History?* (Princeton, NJ: Princeton University Press, 2016).

25 Michael Werner and Bénédicte Zimmermann, "Penser l'histoire croisée : entre empirie et réflexivité," *Annales. Histoire, Sciences Sociales* 58, no. 1 (2003): 5–36.

26 Sanjay Subrahmanyam, "Connected Histories: Notes towards a Reconfiguration of Early Modern Eurasia," *Modern Asian Studies* 31, no. 3 (1997): 735–762.

27 Kapil Raj, *Relocating Modern Science: Circulation and the Construction of Knowledge in South Asia and Europe, 1650-1900* (Basingstoke: Palgrave Macmillan, 2007); and Yves Cohen, "Circulating Localities: The Example of Stalinism in the 1930s," *Kritika: Explorations in Russian and Eurasian History* 11 (2010): 11–45.

28 C. A. Bayly, Sven Beckert, Matthew Connelly, Isabel Hofmeyr, Wendy Kozol, and Patricia Seed, "AHR Conversation: On Transnational History," *American Historical Review* 111, no. 5 (2006): 1441–1464.

29 Pierre-Yves Saunier, "Circulations, connexions et espaces transnationaux," *Genèses* 57, no. 4 (2004): 110–126.
30 For instance: James A. Secord, "Knowledge in Transit," *Isis* 95, no. 4 (2004): 654–672; Simon Schaffer, Lissa Roberts, Kapil Raj, and James Delbourgo, eds., *The Brokered World: Go-Betweens and Global Intelligence, 1770-1820* (Sagamore Beach, MA: Science History Publications, 2009); Kapil Raj, *Relocating Modern Science*; Neil Safier, "Global Knowledge on the Move: Itineraries, Amerindian Narratives, and Deep Histories of Science," *Isis* 101, no. 1 (2010): 133–145; Jorge Cañizares-Esguerra, "On Ignored Global 'Scientific Revolutions'," *Journal of Early Modern History* 21, no. 5 (2017): 420–432; Pablo F. Gómez, *The Experiential Caribbean: Creating Knowledge and Healing in the Early Modern Atlantic* (Chapel Hill, NC: University of North Carolina Press, 2017).
31 Fa-ti Fan, "The Global Turn in the History of Science," *East Asian Science, Technology and Society* 6, no. 2 (2012): 249–258.
32 Dominique Pestre, "Debates in Transnational and Science Studies: A Defence and Illustration of the Virtues of Intellectual Tolerance," *British Journal for the History of Science* 45, no. 3 (2012): 425–442.
33 Stefanie Gänger, "Circulation: Reflections on Circularity, Entity, and Liquidity in the Language of Global History," *Journal of Global History* 12, no. 3 (2017): 303–318.
34 Carlo Ginzburg, "Latitude, Slaves, and the Bible: An Experiment in Microhistory," *Critical Inquiry* 31, no. 3 (2005): 665–683; Linda Colley, *The Ordeal of Elizabeth Marsh: A Woman in World History* (New York, NY: Pantheon Books, 2007); Giovanni Levi, "Microhistoria e Historia Global," *Historia Crítica* 69 (2018): 21–35; John-Paul A. Ghobrial, "Introduction: Seeing the World like a Microhistorian," *Past & Present* 242, Supplement 14 (2019): 1–22; Sarah Easterby-Smith, "Recalcitrant Seeds: Material Culture and the Global History of Science," *Past & Present* 242, Supplement 14 (2019): 215–242.
35 Kapil Raj, "Spaces of Circulation and Empires of Knowledge: Ethnolinguistics and Cartography in Early Colonial India," in *Empires of Knowledge: Scientific Networks in the Early Modern World*, ed. Paula Findlen (New York, NY: Routledge, 2019), 269–294.
36 Latour, *Science in Action*, chap. 6.
37 Heike Jöns, "Centre of Calculation," in *The SAGE Handbook of Geographical Knowledge*, eds. John Agnew and David N. Livingstone (London: SAGE Publications, 2011), 158–170.

Part 1

Knowledge Production on Imperial Landscapes

2 Putting Ships to New Uses

"Floating Gardens" and the Circulation of Knowledge at Sea and on Land, 1790–1800

Jordan Goodman

Introduction

In his path-breaking article of 1996, Richard Sorrenson reminded historians of science in particular that ships were not simply conveyors of "investigators to observe mundane new worlds" but instruments of astronomy and geography in their own right alongside those, such as the telescope, that are viewed quintessentially as such.[1] Sorrenson was absolutely right that ships were or could be used as instruments and, therefore, capable of and indeed enrolled in the process of making knowledge. But, as I will show in this chapter, there was much more to it than that.[2] If, instead of thinking of the ship as an entity, either as a transporter or an instrument, we begin to think about it as an assemblage of actors, spaces, and practices, often on the move but also at its mooring, then it begins to resemble another familiar site of knowledge making – the laboratory. Recently Antony Adler has explored this way of thinking about ships and his contribution is thought-provoking. The drawback, however, is that he focusses on the example of the voyage of HMS *Challenger* (1872–1876), a pioneering, purpose-equipped expedition vessel, constraining his useful discussion about field and laboratory to a pre-designated and designed space.[3]

That ships were the quintessential mobile object; and the[4] primary means by which goods and people moved around the world before the twentieth century is by no means a revelation, but that they were involved in moving nature is less obvious.[5] It is this role and its implications in the making and remaking of knowledge in motion that I wish to explore. While this is not a widely acknowledged phenomenon, I contend that as a topic it straddles two bodies of literature that are well known: the circulation of knowledge and especially that of science on the move; and objects in motion.[6]

Moving living plants by sea over long distances was very risky and consequently not likely undertaken, but the rewards were enormous – the visual and olfactory pleasure was direct and immediate; there was no guessing when and what might sprout. John Ellis in Britain and Henri-Louis Duhamel du Monceau in France had recommended that the best way of ensuring a safe arrival was that the plants should be packed in organic matter and placed in specially designed containers keeping inquisitive shipboard animals at bay while also allowing proper watering and fresh

DOI: 10.4324/9780367814540-3

air to circulate.[7] If amenable to the commission, the ship's commander would share his personal cabin with the plant containers as he did with his log books, tables, charts, and navigational instruments; and he would add the care of living plants, attending to a supply of water and access to light, to the other responsibilities he had. This arrangement worked tolerably well for small volumes: it had little impact on knowledge and its circulation.

Enter Joseph Banks

HMS *Providence* was launched in early July 1791 and though the ship was new the story around it was more than two decades old with origins in Tahiti, the West Indies, and London.[8] At the time of the *Providence*'s launch, Joseph Banks was President of the Royal Society, but in 1769, he was in Tahiti aboard HMS *Endeavour* which was sent to the island to observe and measure the transit of Venus.[9] Banks was not involved in the astronomical work, but instead observed and collected natural history specimens noting especially the breadfruit plant. Banks was one of the first Europeans to see breadfruit growing and to learn about its preparation, but he did not publish anything about it. Others did and by 1775 the virtues of breadfruit – its ease of cultivation and its possibility as an alternative to grain for making bread – were well known. The members of the West India Committee, a powerful parliamentary lobbying group of absentee planters in London, were quick to realize the advantages breadfruit offered them and urged the government to introduce the plant to the British West Indies to provide a staple food for the plantation slave workforce.

Political pressure on the British government to introduce breadfruit renewed vigorously once peace ended the American War of Independence in 1783. In the early part of 1787, William Pitt, the Prime Minister, bowed to the voices and agreed to a government-sponsored voyage to transplant breadfruit to the Caribbean.

Several West Indies residents had contacted Banks after his arrival in 1771 from the *Endeavour* voyage, seeking his opinion on breadfruit plant's botanical characteristics and of the practical issues involved in its transplantation but now he became involved in the project itself. By the end of February 1787, Banks had prepared a draft outline plan for Pitt.[10] Because of his rare experience as a naturalist at sea – he had been on three voyages of exploration over a six-year period – he knew more about ships, particularly about the politics of shipboard space, and naval career practices and the circulation of knowledge, than did most civilians. Banks's advice to Pitt reflected and drew on this experience. Banks began by declaring that this voyage was unprecedented – when ships were involved in botanical transfers, they normally performed this with seeds. Rarely did they carry living plants and certainly not in the volumes envisaged. Transferring breadfruit seeds was not a practical proposition since European naturalists did not know how and when they germinated.

It was going to be a challenging project. Nothing would be taken for granted. "As the Sole Object... in chartering this Vessell ... is to Furnish the West Indian Islands with the Bread Fruit," Banks remarked, "the Master & Crew of her must not think it a grievance to give up the Best part of her accommodations for that purpose."[11] The ship's commander was not immune. "It is necessary therefore," Banks added, "that the Cabbin be appropriated to the Sole purpose of making a Kind of Greenhouse & the *Key of it given to the Custody of the Gardiner* (my emphasis)." Water was no less sacred on ship – it's provision and rationing was normally the commander's responsibility, but here too nothing was the same. Banks continued: "As the plants will Frequently want to be washd [sic] from the Salt dampness Which the Sea air will deposit upon them … a Considerable provision must be made for that purpose … the Gardiner may never be refused the Quantity of Water he may have occasion to demand."

Every bit of free space needed to be filled with plants. This included the quarter-deck, a space usually reserved for the ship's officers, where the potting tubs would be lashed to the sides. The gardener would have to be provided with sufficient canvas to cover the plants when necessary; and the crew could be called upon to help move the heavy tubs periodically from the cabin to the quarterdeck to benefit from sunshine. The commander had a further part to play: no animals, apart from those that provided food, would be allowed on board – cats, dogs, and monkeys were to be excluded; and everything possible had to be done to eradicate rats and cockroaches a feature of shipboard life usually taken for granted. The gardener, therefore, was in charge of everything apart from sailing the ship and the health of its company. For this voyage, Banks personally selected David Nelson, who had sailed as his collector on HMS *Discovery* on Cook's third voyage to the Pacific, and as his assistant, William Brown.[12] The ship was named HMS *Bounty*. William Bligh was given command. Once the alterations to the ship's space had been made according to Banks's specifications, it was possible for the first time to determine its carrying capacity – 800 potting tubs, which could between them carry several thousand plants, were ordered and placed on board.

It all looked set but then Banks intervened and changed the scope of the venture: transplanting the breadfruit would still be its primary objective, but Banks now saw a great opportunity for what he called his "favorite project," that of supplying King George III's garden at Kew with plants from all over the globe. Banks told Nelson that he was to reserve examples of each plant specimen he collected in Tahiti, and elsewhere the ship called at, for Kew. At St Vincent and Jamaica, where the breadfruit would be landed, he was to take on board the plants that Banks had arranged should be assembled for the King.[13] In addition to breadfruit, Nelson was now responsible for other tropical plants new to him.

The *Bounty*, a relatively small ship of 215 tons, left England with forty-six men on 23 December 1787 for the Pacific on a voyage without precedence. Bligh was an excellent sailor, and everything mostly went according to plan. The ship arrived at its anchorage in Matavai Bay on the north coast of Tahiti on 26 October 1788. Bligh knew the place well, having been there in 1777 on Cook's third Pacific voyage.[14] Nelson and Brown got to work immediately. By the beginning of December, the carrying capacity of the cabin had been reached with young breadfruit plants. Bligh ordered the ship's carpenters to make more room on the quarterdeck for a greenhouse for breadfruit and other Tahitian fruit varieties. On 4 April 1789, Bligh was ready to go. The young plants had rooted, and they could now grow as the ship headed to the Caribbean; the ship's vermin had been eradicated. The botanical nursery set sail for the west but then, quite suddenly, on 28 April 1789 the ship was seized by some of the men and Bligh arrested. A few hours after the mutiny, Bligh and eighteen other men, including Nelson, were cast adrift in a launch to find their own way home. More than a thousand plants disappeared from view – they would later end up at the bottom of the sea. The venture was suddenly over. St Vincent and Jamaica would have no breadfruit; and the King would get nothing for his garden.

The Second Breadfruit Voyage

Miraculously, Bligh and most of his men survived the almost 7,000-kilometre journey to Timor and arrived back in England. No one died on the marathon trip but several men, including Nelson, died in Timor. Not only were the plants lost but all of the knowledge that had accumulated about interoceanic horticultural practices also perished. The mutiny did not concern Banks but the loss of plants and knowledge did. Bligh was back in London in March 1790 and his court martial hearing, at which he was pardoned for losing his ship, was held on 22 October 1790. Not long after that, plans were underway for a second attempt with Banks and Bligh in a similar role.[15]

The new ship, HMS *Providence*, was twice the size of the *Bounty* and had room for around 2,000 potting tubs compared to the *Bounty*'s 800.[16] The ship was armed with a marine contingent – no risk of a mutiny again – and the *Providence* would also have a tender.[17] From the plants' point of view, however, there was little difference between the *Providence* and the *Bounty*: the fittings and internal architecture were the same – the captain's cabin was again converted into a nursery for the plants. This was to be an even bigger operation than the *Bounty*'s. Instead of simply transplanting Tahitian plants to the West Indies and collecting examples of tropical plants for Kew, Banks planned for a much more complex botanical interchange, on a scale never before witnessed, including not just Tahiti and the West Indies but also Timor, Cape Town, St Helena, and London. Banks selected two new gardeners, James Wiles and Christopher Smith, who had worked for private gardens in Leeds and London, respectively, where they had learned about caring for tropical plants.[18]

For the outward part of the voyage, Banks purchased pineapples and nectarines from Henry and Hugh Ronalds's nursery.[19] The pineapples were to be a present for the Tahitians, whereas the nectarines were to be exchanged at the Cape for other plants.[20] What else would be moved and where to depended entirely on where Bligh decided to buy the ship's provisions – Madagascar, for example, was mentioned in the instructions as was Mauritius.[21] On the homeward voyage, once Bligh had collected the breadfruit, he was to call at either Timor or Java where he was to replace the dead or dying breadfruit trees with fruit trees from the region – mangosteens, durians, and jackfruits – and also to pick up a type of rice that grew in dry soil. At St Helena, another stop, Bligh was instructed to leave behind some of the breadfruit plants, some of the fruit trees, and some of the rice in sufficient quantities for possible cultivation on the island. In return, he was to take plants from St Helena for the West Indies and for Kew. At the next stop, St Vincent, Bligh was to leave half of the breadfruit trees (reserving for Kew some trees and seeds), in exchange for plants and seeds destined for Kew; and at Jamaica, Bligh was to leave the rest of the breadfruit trees, other plants, and seeds and return to England "bringing with you such Plants, Seeds, etc as the Director [of the botanic garden], or any other Person willing to promote Botany may put on board the ship for his Majesty's said Garden."[22] Bligh was given an advance of £500 in order to purchase any plants he found interesting en route back from Tahiti.[23]

On 2 August 1791, the ships headed off towards warmer climates. The voyage southward through the Atlantic was slow because of contrary winds. The ships arrived at the Cape of Good Hope on 7 November. The plants that Wiles and Smith had brought with them from England had not done well. Many of the nectarine trees had died, though surprisingly the pineapples were unaffected, as were several types of trees. Bligh recommended that fig trees should take the place of the dead nectarine trees: by caring for them, on their route to Tahiti, Bligh hoped that the gardeners would "by their treatment gain some knowledge how to manage the Breadfruit."[24] Wiles and Smith obliged and sought the help of Francis Masson, Kew's first botanical collector who was conveniently at the Cape, in selecting the best varieties and procuring other plants to be potted into the garden's tubs.[25]

On 23 December 1791, the anchors were pulled up and the two ships headed into the Indian Ocean.[26] Bligh was making for Adventure Bay, at the southeast corner of Tasmania for supplies of water and wood – this was his third visit there. The ships arrived in Adventure Bay on 9 February 1792. For Wiles and Smith, it was an opportunity for botanizing. The haul was respectable. They also planted out the botanical specimens they had brought from England and those they had picked up at the Cape. At a lake near the east end of the beach and on a nearby island, Wiles and Smith planted a variety of fruit and wood trees and bushes.[27] Bligh also left behind a cock and two hens – on his first voyage Cook had done the same with hogs.[28]

Bligh kept following the route he had taken with the *Bounty* and headed directly for Tahiti's northern coast, which they reached on 9 April 1792. Bligh was known to many people on the island, and they welcomed him back. He could speak some Tahitian and there were several Tahitians who could speak English, two of them at least, very well. With that and the presents that he had assembled in the form of tools, nails, and other iron products; calico dresses, two hundred yards of chintz, ribbons, shirts, and suits of clothing; knives, hatchets, files, rasps, looking glasses, beads, rings, and combs; not to forget the many plants he had brought with him, from England, the Cape, and Adventure Bay, Bligh felt confident that breadfruit plants would be delivered to him.[29] And so they were. The first consignment of plants, thirty-two in total, arrived on 17 April and were put in the containers they had brought from England.[30]

Wiles and Smith were looking for a spot providing shade and protection from sea breezes, where a permanent structure could be erected to house the plants they had brought with them – and the empty plant tubs awaiting the young breadfruit plants.[31] They found a suitable place on the banks of a small river, 400 meters from Point Venus, under the shade of some enormous breadfruit trees. The ship's carpenters quickly erected a shed, a "House, slightly constructed after the Otehe-itean Manner without a roof, but which they covered at proper Times with Matts made of Cocoa nutt leaves called Pahwahs – the sides they covered with leaves of the Pandanus of Palm." Each day pots were moved from ship to shore and then it was time to get the breadfruit plants. The first method, that of sending Tahitians with empty pots to the breadfruit plantations and then returning to Point Venus with the plants potted, did not work out. Instead, Wiles and Smith found a good supply of loam which they kept in the shed. The plants were brought in from the interior in the morning and were potted in the evening. This process continued for several weeks. Bligh kept a careful count of the number of pots. By early May, the shed was full of 1,000 pots planted with breadfruit. Wiles and Smith looked after them carefully, waiting anxiously for the plants to root so that they could be moved on to the *Providence*. The ship's carpenters, meanwhile, were busy build-ing a greenhouse – this was Bligh's idea, "my own contrivance" – on the ship's quarterdeck which increased the ship's carrying capacity by a third.[32] By mid-June the carpenters had completed the greenhouse. It was built of wood, with a skylight, over which netting could be placed.[33] A railing was fitted to protect the skylight and to enable coverings to be draped over the structure if necessary. The cabin was also prepared, including altering the ports so that fresh air could enter even when they were closed. Bligh was satisfied. "The Plants are doing exceedingly well," he wrote, "which is a peculiar happiness to me."[34]

With the plants on the shore and the ship's space altered accordingly, all was ready apart from the thorough delousing, an operation that lasted several full days. The plants, however, were not ready – they had not yet shown roots. A week went by. Bligh was keen to depart. On Friday 6 July, threatening clouds rolled in. It rained heavily all that day. The following day, the weather returned to its normal state of light winds and mostly clear skies. "My plants," Bligh noted in his log, "have received vast benefit from the Rains, & I hope in ten days they will be fit to be received on board, as I am now anxious about my time." On 14 July, Wiles and Smith gave Bligh the all-clear – the plants had rooted and could now be taken on board. Almost seven hundred pots, many with two plants in them, were moved on the first day and the remaining pots a few days later. The *Providence* dropped six inches into the water because of the weight. On 19 July 1792, final farewells were made. The northwest monsoon, bringing heavy rains and difficult seas, was approaching. By noon, the ships were off for Timor. Bligh happily enumerated the size of the botanic haul from Tahiti. According to his reckoning, there were 1151 vessels filled with breadfruit plants, which contained 2126 individual plants. In addition, there were 508 other plants making a grand total of 2634 plants on the ship.[35]

They arrived in Timor on 2 October 1792. Bligh wanted his stay to be as short as possible. The approaching monsoon, as he put it, "alarms me much."[36] The water cisterns were quickly filled and the breadfruit plants that had died in about 200 pots (a loss of around 20 per cent) were replaced by local plants as Banks had suggested. Banks had anticipated that there would be losses. This did not seem to alarm Bligh or Wiles or Smith. Bligh did not even try to explain it, but Wiles and Smith thought the lack of good, circulating air, in the cabin nursery was the cause.[37] The plants that were in the greenhouse on the deck were hardly affected. This was a lesson, one of several the voyage produced, that would come in useful in the future. Appropriating the commander's cabin and turning it into a plant nursery would never be, apart from the exceptional circumstances of the breadfruit transplant, a viable arrangement. That the plants did better in their own cabin on the quarterdeck was excellent news.[38]

Bligh decided to sail straight through to St Helena in the Atlantic Ocean which they reached on 16 December 1792. "I give you joy of the success of your Plants," Bligh wrote with delight to Banks.[39] It had been more than two months since the ships had left Timor. Bligh had been rightly anxious about his precious cargo and took stock as soon as the ships crossed the Tropic of Capricorn, just to the east of Mauritius. Almost 200 pots of breadfruit plants had perished since leaving Timor. Bligh explained that the plants could not bear the excessive heat during the earlier part of the voyage, especially through the Torres Straits. Those that were near the cabin wall suffered the most. Now that he was in cooler conditions, he did not expect losses at that rate again. Wiles and Smith were also delighted with the voyage so far though they, too, wrote about the losses. They were not alarmed. They repeated that the losses were far greater in the interior cabin than they were in the greenhouse on the deck.[40]

The instructions to Bligh, Wiles, and Smith had specified that when they got to St Helena they were to deliver some of the breadfruit plants and samples of other plants collected in Tahiti and en route to the Atlantic; and, in return, they were to take on board those plants that the Governor wished to present to Kew and those that he and the gardeners felt would be needed and would do well in the West Indies. Banks added that Wiles and Smith should look out for the island's indigenous fern tree several of which he wanted dug up with a ball of local soil, to bring back to England. Wiles and Smith were delighted to see that the breadfruit plants had rooted successfully and that the process of transplanting them into a new habitat went very smoothly. In return for the breadfruit, Wiles and Smith brought on board three fern trees and a collection of plants, including coffee, green tea, almonds, several camellias, and primula, 58 plants in total.[41] Lying at anchor in Jamestown's bay, the *Providence* was quite a sight. Not many ships arriving, certainly no armed naval vessels, would have looked anything like it. "A floating Garden" was how William Doveton, the Secretary to the Council, described it.[42]

The ships were now ready to leave and on 26 December, they began the last leg of their monumental voyage to the Caribbean. Wiles and Smith busily cared for their plants. This was the crucial time. Most days were dry but on one occasion when light rain began to fall and continued to do so for two days, they brought as many plants as they could out of the cabin and onto the deck to benefit from this abundant supply of fresh water.[43] Three weeks after leaving St Helena, they entered the warm waters of the Caribbean, reaching St Vincent, their first port of call, on 23 January 1793. Many plants had died on the crossing: Wiles and Smith observed again that most of the casualties happened in the cabin.[44] Once anchored,

Bligh took another count of his botanical bounty and was pleased that he had 1,200 plants from Adventure Bay, Tahiti, Timor, and St Helena alive.[45] About half of the total plants earmarked for the West Indies were left in the island's botanic garden. Alexander Anderson, its superintendent, had prepared a collection of plants to form part of the West Indian collection for Kew. These were put on the *Providence*, taking the place of the plants that had been left behind. Bligh was relieved to have come this far.[46]

The voyage westward across the Caribbean Sea to Jamaica was quick. Just a week later, the ships anchored in Port Royal, Jamaica, on 5 February 1793. The authorities were not expecting Bligh so soon and were not prepared but not long after they agreed what to do with the breadfruit plants.[47] Meanwhile, Wiles, Smith, and the island's botanical personnel began preparing the plants that had been put aside for the King's garden at Kew. Pots, tubs, and boxes full of hundreds of plants were put on board filling the room in the interior and in the greenhouse on the deck, which had previously housed the plants for St Vincent and Jamaica. Altogether, Bligh recorded that there were 796 containers holding 1,283 plants bound for London. Two days later, on 4 June 1793, Bligh weighed anchor and set sail for home. The "floating garden" was at sea once more. Wiles had decided to remain behind, a possibility that he and Banks had agreed to in London. With him he had a Tahitian man, Pappo, who had been helping Wiles and Smith collect breadfruit plants and other fruit trees during the time the ships were in Tahiti, and who had stowed away on the *Providence*. Wiles and Smith explained to Bligh how valuable he had been to them on the island and that he would continue to be useful in Jamaica to carry on this work and impart what he knew there.[48] Pappo taught Wiles many valuable lessons on how to look after the Tahitian plants but he soon died despite being inoculated against smallpox.[49] In Wiles's absence, Smith was left in charge of the tropical plants in the cabin and on the quarterdeck as the ships sailed eastward across the Atlantic to London. He was not only bringing plants back for Kew but also an unsurpassed wealth of knowledge about how to move living plants across the globe.

Completing the Circle: The Voyage of the *Royal Admiral*

The *Providence* arrived in Deptford on 7 August 1793. The pots, tubs, and boxes, containing almost two thousand plants, the single largest shipment of plants from the West Indies to London yet, were transferred onto lighter craft for their trip upstream to Kew.[50] In addition, there were four breadfruit plants and twenty-eight other plants representing sixteen specimens, which had travelled more than half way across the world and which had survived aboard ship for more than a year.[51] Smith accompanied the plants to Kew to care for them in this critical moment of transplantation.[52] In early September 1792, the Court of Directors of the East India Company in London were informed that there was a vacancy for a gardener in the Calcutta Botanic Garden.[53] The Court of Directors turned to Banks for a recommendation and, without hesitating, he suggested Smith.[54]

Banks would not let a botanical opportunity slip by. Smith had unparalleled knowledge of tropical and temperate plants, and how to keep them alive at sea over long distances. Having Smith travel to Calcutta on an East India Company ship as a passenger would have been, to Banks, a waste of the application of knowledge. The King's garden at Kew had few plants from Bengal and its environs. These were being grown in the East India Company's Calcutta Botanic Garden. Colonel Robert Kyd, its first superintendent, had sent only a few examples to Kew and those mostly died on the voyage to London.[55] Kyd's successor, William Roxburgh, an East India Company botanist who had been corresponding with Banks for fifteen years, was going to be more forthcoming.[56]

Banks wanted a two-way plant exchange between Calcutta and London but he had a problem: Smith was going to Calcutta for some time, possibly permanently. Banks needed someone else to accompany Smith to Calcutta and return with the chosen collection, preferably someone who was also familiar with Kew. In May 1794, Banks contacted William Aiton, Kew's head gardener, for help in finding such a person. Not long after, Aiton wrote back that Peter Good, who had been working at Kew for two years, was the man.[57] Banks agreed: he could now outline his plan, one which would be mutually advantageous to Kew and the Calcutta Botanic Garden, to William Devaynes, Chairman of the East India Company. It went like this: Peter Good would go out with Smith and return with the plants for Kew. To give him the practical experience to take best care of the plants on the return journey, Banks suggested that the ship should take out to Calcutta specimens of fruit trees, many of which Colonel Kyd had specified would do well, and other useful plants, some that the *Providence* had returned with, and that these should be entrusted to Smith's care.[58]

The fruit trees and other plants, Banks advised, should be housed in a plant cabin erected on the ship's quarterdeck, which could also be used for the plants on the return voyage. Banks had already had several good experiences with these glazed structures, the first of which he had designed and had built on a Royal Navy vessel in 1789, but this would be the first time one would be built on an East India Company ship.[59] Good's main responsibility was to Smith and to the plants. On the way to Calcutta, he was to follow all of Smith's directions and to learn from him about how to care for plants at sea, and particularly how a plant cabin worked. While the *Providence* had not been equipped with a glazed plant cabin on its deck, the arrangement of the ship's garden and the alterations made to the deck by Bligh acted very much like such a cabin. Smith (and Banks) had also learned during the voyage of the *Providence* that plants on the quarterdeck did better than those in the interior cabin. On the homeward voyage, the plants for the King, which would be kept in the plant cabin, would be entirely Good's responsibility. Banks was leaving nothing to chance. As soon as he had given Good his instructions, Banks wrote to Devaynes that the ship's commander was critical to protecting the plants and that it might be a good idea to add these to his usual sailing instructions.

At over 900 tons, the *Royal Admiral* was one of the largest ships in the East India Company's fleet and had recently completed its sixth voyage to and from India and China. By the end of July 1794, almost everything was in place to bring Banks's grand scheme for the transfer of plants to and from Calcutta to fruition. The gardeners had been chosen and given their orders; the plants assembled for the outward voyage; and the ship and its glazed plant cabin readied for departure. The ship's commander was Captain Essex Henry Bond. The plants could not be in better hands on both legs of the voyage, Bond assured Banks. The weather, which he felt would be kind to the voyage at this time of the year, was sure to play its part in ensuring success, but in addition, Bond told Banks that he had already had a successful experience transporting plants long distances. On a previous voyage, he remarked, he had taken on board several dozen fruit trees at the Cape on his way to Sydney to land convicts there and these, he had learned, "were doing exceedingly well in the colony."[60]

The *Royal Admiral* left Portsmouth harbor on 14 August 1794. On 4 September, it arrived in Madeira. A day later, Smith wrote to Banks. "It is with pleasure I tell you," he wrote, "that my little Vegetable family continues in Luxuriant health partly owing to their own good Stamina and somewhat to my Paternal attention."[61] When the ship got to Calcutta, on 25 February, Smith reported with not a little pride that the plants were in excellent condition. The shipment of three hundred and nine separate plants represented, as Smith put it, "the largest Collection of plants that ever has been introduc'd from Europe into India."[62] Captain Bond was as good as his word having "paid great attention to the plants ... and always ready to supply ... everything that was necessary for their preservation."[63] Many of the plants were entirely new to Bengal – Banks had chosen wisely.[64] Smith had received instructions from Banks that when he was at Madeira and the Cape he should augment the shipboard collection of fruit trees by purchasing local varieties which, in his estimation, would succeed in Calcutta. For this, Captain Bond advanced him with the necessary funds – a little over £30.[65] Madeira supplied grape vines as did the Cape in addition to a variety of vegetable plants and many flowers.[66]

In Calcutta, Smith quickly began moving the plants from the ship to the garden. And as soon as that was done, he began collecting the plants for the homeward voyage on 4 March. Six weeks later the collection was on board the *Royal Admiral*. The ship only stayed in Calcutta as long as it took to unload one set of plants and replace them with the Indian specimens. On 26 April, Christopher Smith drew up his final list of 375 plants, some singles, others duplicates, all of which, Smith noted, "I have potted with my own hands."[67] He was clearly pleased with his collection. He had annotated an earlier version of his list; one plant, which had been named after Sir William Jones, the jurist, orientalist, and founder of the Asiatic Society of Bengal, Smith described: "There is no plant exceeds this for beauty when in flower"; another Indian plant, Smith simply called "glorious."[68]

The *Royal Admiral* now began its homeward voyage. The ship sailed first to Madras where Sir Paul Jodrell, physician to the Nawab of Arcot, donated a large cinnamon tree as a present to the King.[69] Trincomalee, on Sri Lanka's eastern coast, was the ship's next stop and there it remained for the second half of August. Almost twenty plants were collected.[70] It was time to head back to England which Bond did after a short layover in St Helena. When the plants finally arrived at Kew in May 1796, Banks wrote to William Roxburgh in Calcutta and with no little pride remarked that "next to the Collection which Captain Bligh brought home it is the largest addition Kew Gardens have ever received at one time."[71] Christopher Smith remained in the East Indies, as expected. In 1805, and no doubt recommended by Banks, he became the superintendent of the East India Company's first botanic garden in Penang but unfortunately did not live long enough to do anything other than get it started – he died in 1807.[72] Soon, in order to keep the knowledge of moving living plants across the world's oceans fully up-to-date, Peter Good would be on yet another voyage of exploration taking him even further afield to Australia with Matthew Flinders on a circumnavigation of Australia, on another "floating garden."[73]

Conclusion

Banks clearly liked using glazed plant cabins to house living plants as they moved across the world's seas on the deck of ships. When the *Royal Admiral* was fitted out in 1794 in this manner for its voyage to and from Calcutta, Banks had already used the plant cabin on a naval voyage to Australia and currently on a naval voyage of exploration in the Pacific and the northwest coast of America. He would in future be doing the same with a present of living plants from Kew for the Empress of Russia's garden in St. Petersburg; and a series of almost ten voyages to and from Canton between 1803 and 1811.[74] This technology, which had never been tried before to transport living plants, was only part of a dynamic package of knowledge which was put on board ship in the same way as were the people and objects that made the ship sail. There were also gardeners and the ship's commander and, in some of the later voyages, even passengers with horticultural experience. Banks ensured that this knowledge of moving plants would circulate between voyages in the same way as maritime knowledge circulated through the career structure and pattern on both naval and commercial ships.

Notes

1 Richard Sorrenson, "The Ship as a Scientific Instrument in the Eighteenth Century," *Osiris* 11 (1996): 221–236.
2 See Martin Dusinberre and Roland Wenzlhuemer, "Editorial – Being in Transit: Ships and Global Incompatibilities," *Journal of Global History* 11 (2016): 155–162 for a very wide view of the ship, as a "site of history," "a temporal and spatial mediator" (p. 158). They align their views with those expressed in Simon Schaffer, Lissa Roberts, Kapil Raj, and James Delbourgo, eds, *The Brokered World: Go-Betweens and Global Intelligence, 1770–1820* (Sagamore Beach, MA: Science History Publications, 2009).
3 Antony Adler, "The Ship as Laboratory: Making Space for Field Science at Sea," *Journal of the History of Biology* 47 (2014): 333–362 and Antony Adler, *Neptune's Laboratory: Fantasy, Fear, and Science at Sea* (Cambridge, MA: Harvard University Press, 2019). See also an even more recent analysis of the *Challenger's* voyage, in which the ship's time at rest is given as much concern as its time at sea, a point which is rarely made in the literature – Erika Lynn Jones, *Making the Ocean Visible: Science and Mobility on the* Challenger *Expedition, 1872–1895* (PhD unpublished thesis, University College London, 2019). See also Kimberley Peters, "Drifting: Towards Mobilities at Sea," *Transactions of the Institute of British Geographers* 40 (2015): 262–273.
4 For ships as mobile entities and sites of practice, see Ofer Gal and Yi Zheng, eds., *Motion and Knowledge in the Changing Early Modern World* (Dordrecht: Springer, 2014);

Anyaa Anim-Addo, William Hasty, and Kimberley Peters, "The Mobilities of Ships and Shipped Mobilities," *Mobilities* 9 (2014): 337–349; Sarah Louise Millar, "Science at Sea: Soundings and Instrumental Knowledge in British Polar Expedition Narratives, c.1818–1848," *Journal of Historical Geography* 42 (2013): 77–87; and Michael S. Reidy and Helen Rozwadowski, "The Spaces In Between: Science, Ocean, Empire," *Isis* 105 (2014): 338–351. This body of literature spills easily into another, that of the sea as the object of historical study. For this large topic, see David Lambert, Luciana Martins, and Miles Ogborn, "Currents, Visions and Voyages: Historical Geographies of the Sea," *Journal of Historical Geography* 32 (2006): 479–493; Michael North, "Connected Seas," *History Compass* 16/12 (2018), n.p., especially the references therein; and André Dodeman and Nancy Pedri, *Negotiating Waters: Seas, Oceans, and Passageways in the Colonial and Postcolonial Anglophone World* (Wilmington, DE: Vernon Press, 2020).

5 See Alan Bewell, *Natures in Translation: Romanticism and Colonial Natural History* (Baltimore, MD: Johns Hopkins University Press, 2017) and Alan Bewell, "Traveling Natures," *Nineteenth–Century Contexts* 29 (2007): 89–110.

6 A good place to enter into the literature on the circulation of knowledge is with the published works of Kapil Raj – see, especially, *Relocating Modern Science: Circulation and the Construction of Scientific Knowledge in South Asia and Europe, 1650–1900* (Basingstoke: Palgrave Macmillan, 2007); "Beyond Postcolonialism … and Postpositivism: Circulation and the Global History of Science," *Isis* 104 (2013): 337–347; "Go-Betweens, Travelers, and Cultural Translators," in Bernard Lightman, ed., *A Companion to the History of Science* (Chichester: Wiley, 2016), 39–57; "Networks of Knowledge, or Spaces of Circulation? The Birth of British Cartography in Colonial South Asia in the Late Eighteenth Century," *Global Intellectual History* 2 (2017): 49–66; and "Spaces of Circulation: Ethnolinguistics and Cartography in Early Colonial India," in Paula Findlen, ed., *Empires of Knowledge: Scientific Networks in the Early Modern World* (London: Routledge, 2019), 269–293. An example of one subset of this literature, focussed on expeditions and explorations, includes Kapil Raj, "18th–Century Pacific Voyages of Discovery, 'Big Science', and the Shaping of an European Scientific and Technological Culture," *History and Technology* 17 (2000): 79–98, Marianne Klemun and Ulrike Spring, eds. *Expeditions as Experiments: Practising Observation and Documentation* (London: Palgrave Macmillan, 2016); Nicolas Dew, "Scientific Travel in the Atlantic World: The French Expedition to Gorée and the Antilles, 1681–1683," *The British Journal for the History of Science* 43 (2010): 1–17; and Jean Fornasiero and John West-Sooby, "The Narrative Interruptions of Science: The Baudin Expedition to Australia (1800–1804)," *Forum for Modern language Studies* 49 (2013): 457–471. The literature on studying objects in motion is large and growing. A good place to start is Anne Gerritsen and Giorgio Riello, eds. *The Global Lives of Things: The Material Culture of Connections in the Early Modern World* (London: Routledge, 2016) and Meredith Martin and Daniela Bleichmar, eds., "Objects in Motion in the Early Modern World," *Art History* 38 (2015): 605–619.

7 The literature on the history of moving plants by sea is limited but see the following: Christopher M. Parsons and Kathleen S. Murphy, "Ecosystems Under Sail: Specimen Transport in the Eighteenth–Century French and British Atlantics," *Early American Studies*, 10 (2012): 503–529, and the references there to Ellis and Duhamel du Monceau; Nigel Rigby, "The Politics and Pragmatics of Seaborne Plant Transportation," in Margarette Lincoln, ed., *Science and Exploration in the Pacific: European Voyages to the Southern Oceans in the Eighteenth Century* (Woodbridge: Boydell Press, 1998), 81–100; Yves–Marie Allain, *Voyages et survie des plantes au temps de la voile* (Marly-le-Roi: Editions Champflour, 2000); and *Moved Natural Objects. Spaces in Between*, Special issue of *HOST: Journal of History of Science and Technology* 5 (2012), especially the two articles by Marianne Klemun, "Introduction: 'Moved' Natural Objects – 'Spaces in Between'," 9–16 and "Live Plants on the Way: Ship, Island, Botanical

Garden, Paradise and Container as Systemic Flexible Connected Spaces in Between," 30–48. See also Mark Laird and Karen Bridgman, "American Roots: Techniques of Plant Transportation in the Early Atlantic World," in Pamela H. Smith, Amy R. W. Meyers, and Harold J. Cook, eds., *Ways of Making and Knowing: The Material Culture of Empirical Knowledge* (Ann Arbor, MI: University of Michigan Press, 2014), 164–193 and Charles E. Nelson, "From Tubs to Flying Boats: Episodes in Transporting Living Plants," in Arthur MacGregor, ed., *Naturalists in the Field: Collecting, Recording and Preserving the Natural World from the Fifteenth to the Twenty–First Century* (Leiden: Brill, 2018), 578–606. A one-page contemporary statement on the best methods for moving plants can be found in [John Fothergill], "Directions for Taking Up Plants and Shrubs, and Conveying Them by Sea" (1790?). The plant containers advocated by Ellis and Duhamel du Monceau should not be confused with the self-contained glazed Wardian cases, named after their inventor Nathaniel Bagshaw Ward, which became widely used to transport living plants in the 1840s and beyond – see Margaret Flanders Darby, "UnNatural History: Ward's Glass Cases," *Victorian Literature and Culture* 35 (2007): 635–647, Luke Keogh, "The Wardian Case: Environmental Histories of a Box for Moving Plants," *Environment and History* 25 (2019): 219–244, and Stuart McCook, "'Squares of Tropic Summer': The Wardian Case, Victorian Horticulture, and the Logistics of Global Plant Transfers, 1770–1910," in Patrick Manning and Daniel Rood, eds., *Global Scientific Practice in an Age of Revolutions, 1750–1850* (Pittsburgh, PA: University of Pittsburgh Press, 2016), 199–215.

8 The full story is in my *Planting the World: Joseph Banks and his Collectors – An Adventurous History of Botany* (London: William Collins/HarperCollins, 2020), chapter 8. See also Juliane Braun, "Bioprospecting Breadfruit: Imperial Botany, Transoceanic Relations, and the Politics of Translation," *Early American Literature* 54 (2019): 643–671.

9 Harry Woolf, *The Transits of Venus: A Study of Eighteenth–Century Science* (Princeton, NJ: Princeton University Press, 1959) and Andrea Wulf, *Chasing Venus: The Race to Measure the Heavens* (London: William Heinemann, 2012).

10 Banks to Pitt, [February 1787], in Neil Chambers, ed., *The Indian and Pacific Correspondence of Sir Joseph Banks* (London: Pickering and Chatto, 2009), vol. 2, document 114.

11 This quote, and those in the following section, are from Banks to Pitt, [February 1787], Chambers, *The Indian and Pacific Correspondence*, vol. 2, document 114.

12 Little more is known about him.

13 Banks to Nelson, 20 August 1787, Chambers, *The Indian and Pacific Correspondence*, vol. 2, document 150.

14 See Anne Salmond, *Bligh: William Bligh in the South Seas* (Berkeley, CA: University of California Press, 2011).

15 See Royal Botanic Gardens Kew Archives, JBK/1/6, f. 182. The letter is undated and misbound in a later volume, but from internal evidence had to be before 15 December 1790. Two days later, Banks wrote to William Eden, Britain's Ambassador to the Netherlands, that the King had already ordered a ship to return to Tahiti for the breadfruit plants – Banks to Eden, 17 December 1790, British Library, Add MS 34434, 335–336. See also Yonge to Banks, n.d., in Chambers, *The Indian and Pacific Correspondence of Sir Joseph Banks*, (London: Pickering and Chatto, 2011), vol. 4, document 298.

16 The plans for the *Providence* are in National Maritime Museum, ZAZ6585-6586 and the *Bounty's* are in ZAZ6664-6668 and ZAZ7848.

17 The ship carried fifteen marines to the rank of sergeant – National Archives (Kew), ADM 2/267.

18 William Brown had stayed with the mutineers.

19 See 29 May 1792, Bligh's log at *www.fatefulvoyage.com/providenceBligh/index.html* (accessed 20 August 2018). It is also transcribed on the State Library of Queensland's website and has been published as a facsimile by Genesis Publications, 1976. The invoice for the pineapples is in Sutro Library, Sir Joseph Banks Collection, G 1:18. For pineapples, see Johanna M. Gohmann, "Colonizing through Clay: A Case Study of the

Pineapple in British Material Culture," *Eighteenth–Century Fiction* 31 (2018): 143–161 and the references to works by Levitt, Beauman, and Okihoro, respectively: see also E. J. Cole, *The Cultural History of Exotic Fruits in England 1650–1820* (unpublished PhD dissertation, University of Cambridge, 2006). For Ronalds nursery, see Beverley F. Ronalds, "Ronalds Nurserymen in Brentford and Beyond," *Garden History* 45 (2017): 82–100.

20 National Archives (Kew), HO 28/62 and Sutro Library, Sir Joseph Banks Collection, G 1:18.
21 See 18 April 1791, National Archives (Kew), HO 28/8, 62–63.
22 15 July 1791, National Archives (Kew), ADM 2/121, 475–480.
23 18 April 1791, National Archives (Kew), HO 28/8, 62–63.
24 Bligh to Banks, 24 November 1791, in Chambers,*The Indian and Pacific Correspondence of Sir Joseph Banks* (London: Pickering and Chatto, 2010), vol. 3, document 214.
25 Wiles and Smith to Banks, 28 November 1791, Chambers, *The Indian and Pacific Correspondence*, vol. 3, document 217 and Bligh to Banks, 17 December 1791, Chambers, *The Indian and Pacific Correspondence*, vol. 3, document 220.
26 Bligh hoped to leave by 6 December but was too ill to proceed – Bligh to Banks, 17, 18 December 1791, Chambers*, The Indian and Pacific Correspondence*, vol. 3, documents 220, 221.
27 Bligh, Remarks, 10–11 February 1792 at *www.fatefulvoyage.com/providenceBligh/index.html* (accessed 20 August 2018).
28 Bligh, Remarks, 22 February 1792, at *www.fatefulvoyage.com/providenceBligh/index.html* (accessed 20 August 2018).
29 The list of presents, which Bligh chose, is in National Archives (Kew), ADM 1/1507.
30 Unless otherwise indicated, all of the following is taken from Bligh's log at *www.fatefulvoyage.com/providenceBligh/index.html* (accessed 20 August 2018).
31 Wiles and Smith to Banks, 17 December 1792, Chambers, *The Indian and Pacific Correspondence*, vol. 4, document 7.
32 Flinder's log at *www.fatefulvoyage.com/providenceFlinders/index.html* (accessed 12 August 2018) and Bligh to Banks, 16 December 1792, Chambers, *The Indian and Pacific Correspondence*, vol. 4, document 6.
33 This and the following detail comes from 1st Lieutenant Francis Bond's log of 24 April and 18 June 1792, at *www.fatefulvoyage.com/providenceBond/index.html* (accessed 17 August 2018).
34 Entry for 18 June 1792 at *www.fatefulvoyage.com/providenceBligh/index.html* (accessed 20 August 2018).
35 Wiles and Smith's figures were significantly lower: they counted 1686 breadfruit plants and 310 other botanical specimens – Wiles and Smith to Banks, 17 December 1792, Chambers, *The Indian and Pacific Correspondence*, vol. 4, document 7.
36 Bligh to Banks, 2 October 1792, Chambers, *The Indian and Pacific Correspondence*, vol. 3, document 313.
37 Wiles and Smith to Banks, 17 December 1792, Chambers, *The Indian and Pacific Correspondence*, vol. 4, document 7.
38 Wiles and Smith to Banks, 17 December 1792, Chambers, *The Indian and Pacific Correspondence*, vol. 4, document 7.
39 Bligh to Banks, 16 December 1792, Chambers, *The Indian and Pacific Correspondence*, vol. 4, document 6.
40 Wiles and Smith to Banks, 17 December 1792, Chambers, *The Indian and Pacific Correspondence*, vol. 4, document 7.
41 Wiles and Smith to Banks, 17 December 1792, Chambers, *The Indian and Pacific Correspondence*, vol. 4, document 7 and Wiles to Banks, 22 January 1793, Chambers, *The Indian and Pacific Correspondence*, vol. 4, document 21.
42 The phrase appears in the 19–26 December entry in Bligh's log at *www.fatefulvoyage.com/providenceBligh/index.html* (accessed 20 August 2018).

43 Flinders's log for 9 January 1793, at *www.fatefulvoyage.com/providenceFlinders/index.html* (accessed 12 August 2018)

44 Wiles to Banks, 22 January 1793, Chambers, *The Indian and Pacific Correspondence*, vol. 4, document 21.

45 Bligh to Banks, 23 January 1793, Chambers, *The Indian and Pacific Correspondence*, vol. 4, document 23.

46 Bligh to Banks, 27 January 1793, Chambers, *The Indian and Pacific Correspondence*, vol. 4, document 25.

47 Most of the plants were held on deposit in the main botanic garden at Bath and some at Spring Garden, once the property of Hinton East, a friend of Banks's and an early advocate of the breadfruit exchange, who had died in the previous year; some, which had already been earmarked, were delivered personally by Bligh in his two ships. See Bligh's log for 11 January and 6–10 February 1793, at *www.fatefulvoyage.com/providenceBligh/index.html* (accessed 20 August 2018) and Richard B. Sheridan, "Captain Bligh, the Breadfruit, and the Botanic Gardens of Jamaica," *Journal of Caribbean History* 23 (1989): 28–50.

48 Bligh's log for 18 July 1792, at *www.fatefulvoyage.com/providenceBligh/index.html* (accessed 20 August 2018).

49 Wiles continued to write to Banks once the ships departed for London: Sutro Library, Sir Joseph Banks Collection, BF 1:9–20. On Pappo's illness see Wiles to Banks, 16 October 1793, in Chambers, *The Indian and Pacific Correspondence*, vol. 4, document 87; and about his death, Wiles to Banks, October 1793, in Chambers, *The Indian and Pacific Correspondence*, vol. 4, document 91.

50 Smith, "Plants on Board H.M.S. Providence," State Library of New South Wales, Sir Joseph Banks Papers, Series 52.16. See Smith to Banks, 1 August 1793, Chambers, *The Indian and Pacific Correspondence*, vol. 4, document 69; Dancer to Banks, 10 January 1793, Chambers, *The Indian and Pacific Correspondence*, vol. 4, document 17; Dancer to Banks, 3 April 1793, Chambers, *The Indian and Pacific Correspondence*, vol. 4, document 45; Aiton to Banks, 10 August 1793, Chambers, *The Indian and Pacific Correspondence*, vol. 4, document 75; and Royal Botanic Gardens Kew Archives, Record Book 1793–1809, 10–16.

51 Smith, "Plants on Board H.M.S. Providence," State Library of New South Wales, Series 52.16.

52 Banks to Nepean, 15 December 1793, in Chambers, *The Indian and Pacific Correspondence*, vol. 4, document 104. Many of the plants were still prospering almost twenty years later – see Dulcie Powell, "The Voyage of the Plant Nursery, H.M.S. *Providence*, 1791–1793," *Economic Botany* 31 (1977): 428–431.

53 "Public Letter from Bengal," 3 September 1792, Asiatic Society of Bengal, Banks MSS, letter 10 (microfilm copy deposited in Banks Archive Project, Nottingham Trent University). Bengal Public Consultations, 23 October 1793, Natural History Museum, DTC, VI, 154.

54 This correspondence can be found in Asiatic Society of Bengal, Banks MSS. See also Banks to Ramsay, 14 May 1793, Chambers, *The Indian and Pacific Correspondence*, vol. 4, document 56. Smith's appointment was ratified near the end of December – Court of Directors to Bengal, 2 July 1794, British Library, IOR, E/4/641.

55 "Plants received from Robert Kyd," June 1790, in Chambers, *The Indian and Pacific Correspondence*, vol. 3, document 93. Kyd began the garden in 1787 with Banks's support – see Adrian P. Thomas, "The Establishment of the Calcutta Botanic Garden: Plant Transfer, Science and the East India Company, 1786–1806," *Journal of the Royal Asiatic Society* 16 (2006): 165–177.

56 Roxburgh to Banks, 1 December 1793, Chambers, *The Indian and Pacific Correspondence*, vol. 4, document 98. For Roxburgh see Tim Robinson, *William Roxburgh: The Founding Father of Indian Botany* (Chichester: Phillimore, 2008).

57 Aiton to Banks, 29 May 1794, State Library of New South Wales, Sir Joseph Banks Papers, Series 20.04.

58   Kyd to Banks, 21 November 1791, Chambers, *The Indian and Pacific Correspondence*, vol. 3, document 213 and Kyd to Court of Directors, 13 August 1787, Chambers, *The Indian and Pacific Correspondence*, vol. 2, document 258 (p. 365).
59   Banks to Devaynes, 3 June 1794, State Library of New South Wales, Sir Joseph Banks Papers, Series 17.01.
60   Bond to Banks, 2 August 1794, Chambers, *The Indian and Pacific Correspondence*, vol. 4, document 135.
61   Smith to Banks, 5 September 1794, Chambers, *The Indian and Pacific Correspondence*, vol. 4, document 139.
62   Smith to Banks, 12 May 1795, Chambers, *The Indian and Pacific Correspondence*, vol. 4, document 184. The plant list of September 1795 is in State Library of New South Wales, Sir Joseph Banks Papers, Series 16.11.
63   Smith to Banks, 12 May 1795, Chambers, *The Indian and Pacific Correspondence*, vol. 4, document 184.
64   Roxburgh to Banks, 25 April 1795, Chambers, *The Indian and Pacific Correspondence*, vol. 4, document 177.
65   Bond to Sir John Shore, 23 April 1795, British Library, IOR, P/4/34, p.519.
66   Smith to Banks, 12 May 1795, Chambers, *The Indian and Pacific Correspondence*, vol. 4, document 184 and State Library of New South Wales, Sir Joseph Banks Papers, Series 16.11. At the Cape, Smith was assisted by Francis Masson, who was still there, and Colonel Robert Jacob Gordon, an experienced botanist and gardener – Patrick Cullinan, *Robert Jacob Gordon 1743–1795: The Man and His Travels at the Cape* (Cape Town: Struik Winchester, 1992).
67   State Library of New South Wales, Sir Joseph Banks Papers, Series 16.13.
68   State Library of New South Wales, Sir Joseph Banks Papers, Series 16.14. For more on Jones, see Kapil Raj, "Refashioning Civilities, Engineering Trust: William Jones, Indian Intermediaries, and the Production of Reliable Legal Knowledge in Late–Eighteenth–Century Bengal," in Kapil Raj, ed., *Relocating Modern Science*, 95–138.
69   State Library of New South Wales, Sir Joseph Banks Papers, Series 16.12. King George III recommended Jodrell for this post in 1787.
70   State Library of New South Wales, Sir Joseph Banks Papers, Series 16.12.
71   Banks to Roxburgh, 29 May 1796, Chambers, *The Indian and Pacific Correspondence*, vol. 4, document 252; and Banks to Smith, 26 May 1796, Chambers, *The Indian and Pacific Correspondence*, vol. 4, document 251. Another plant list is in Royal Botanic Gardens Kew Archives, Record Book 1793–1809, 157–160.
72   Smith died either in Penang or en route to England from there – see Marcus Langdon, *Penang: The Fourth Presidency of India 1805–1830, Volume Two: Fire, Spice & Edifice* (Penang: George Town World Heritage, 2015): Book Five.
73   See my *Planting the World*, chapters 17, 18 for this story.
74   See my *Planting the World*.

3 Regional Knowledge in the Empire

Tobacco Cultivation during the Napoleonic Era

Alexander van Wickeren

Introduction

Rhenish attempts to substitute colonial tobacco in the early nineteenth century by fostered improvement of local leaf cultivation provide so far unconsidered evidence for the regionality of knowledge circulation in the Napoleonic Empire, an era that was uniquely shaped by the spread of modern state administration and nearly 20 years of constant warfare.[1] Around 1800, European and other tobacco cultivation areas became a relevant substitute for raw tobacco from the Americas. Agricultural tobacco research reached a first peak during the Napoleonic era as part of the larger "Agricultural Enlightenment," intensifying the collection and production of "useful knowledge" for the economization of the environment since the mid-eighteenth century.[2] In the years between 1810 and the dissolution of the Napoleonic Empire in 1814, civil servants and enlightened savants in the tobacco-growing regions of the Rhineland, which were formally or informally dominated by France, advocated for tobacco-specific cultivation manuals.[3] Reformers in the Alsace, the northern Rhineland, and the southern Rhine areas of the Palatinate and Baden oriented botanically – in addition to various practical aspects, reaching from seed sowing to harvest – and connected on regional scale.

This chapter shows a specific case of what Kapil Raj has called the "geography" of "spaces of circulation";[4] frameworks for knowledge production for which the rising modern state, with its newly organized administrations and trained public officials, provided a central infrastructure.[5] We know much about how European imperial expansion enabled new relationships between apparent centers and peripheries. Looking at a *Knowing Empire* in Europe offers the fascinating possibility of linking questions about circulation, knowledge, and empire in new ways to research traditions that have examined the genesis of modern science in regard to its spatial organization.[6] Studies have, however, focused primarily on the nationalization of the sciences, i.e., the new appearance of academies organized according to national criteria, networks of scholars oriented around imagined boundaries of national communities, and the emergence of scientific "national styles."[7]

This chapter explores the comparatively less explored regionality in modern sciences as an effect of imperial conquest and state penetration in new land acquisitions within Europe.[8] In many cases, regional spaces of circulation of knowledge

DOI: 10.4324/9780367814540-4

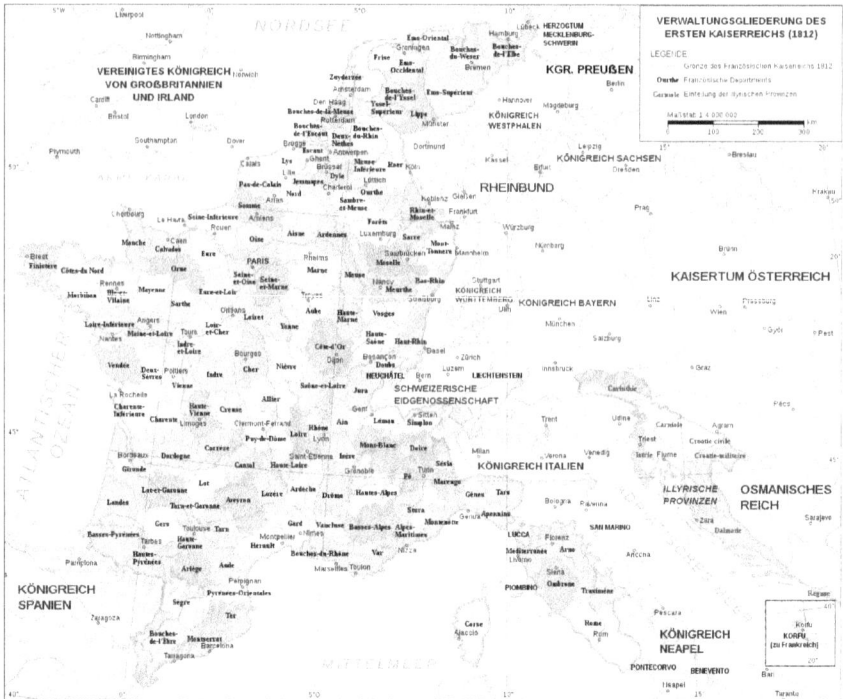

Figure 3.1 Map of the administrative divisions of the First French Empire.

Source: Wikimedia Commons.

overlapped with spaces of identity and loyalty, which, far from fading away as nationalization progressed, rather became more contoured.[9] In other cases, as historians have shown, the regionality of knowledge can rather be understood as a network and connection space above the local, such as the city, and below the nation, which produced enough cohesion to produce similar opinions or attitudes.[10]

Studies of tobacco research have emphasized the importance of political contexts, including imperial rivalry, for the emergence of new expertise, yet rarely from a spatial perspective.[11] War and trade obstacles around 1800 shifted the view to possible import substitutions of apparent "colonial goods," such as tobacco, which in the early modern period were almost exclusively sourced from the colonies in the Americas.[12] Events such as the North American independence wars of the 1770s and 1780s, the official ban on the import of Britain's colonial goods with the Continental Blockade (1806–1813), or the establishment of the state tobacco monopoly in the French Empire in the years after 1810 fostered a "second globalization" of tobacco, as tobacco cultivation expanded into new territories beyond the older cultivation areas in the Atlantic region. Pushed by an increase in tobacco consumption,[13] longer-established cultivation areas became more commercialized, trading tobacco transregionally, sometimes even on global scale.[14]

The extent to which imperial expansion favored regional spaces of knowledge circulations during tobacco's "second globalization" has not been investigated so far. This chapter focuses on the Rhineland as an example. The first section of this chapter explores how the state tobacco monopoly administration in the French Empire became a central factor in the regional exchange among Rhenish experts. The second part is devoted to the specific regional research style of tobacco researchers. The third part then discusses the inclusion and exclusion effects of patronage relations for tobacco research's regionalization. Fourthly, the chapter discusses the passivity of the Parisian central tobacco monopoly administration as a constitutive factor for Rhenish tobacco improvement. Fifthly, and finally, the chapter asks to what extent the rise of national emotions transformed the regional geographies of knowledge circulation.

Regional Circulation

In the Département du Bas-Rhin, the northern part of the Alsace, the position as public "Inspecteur de la culture du tabac"[15] and supervisor of the state-run botanical tobacco research gardens in Schlestadt, Obernai, and Strasbourg[16] had been left to Johan Nepomuk Schwerz who had gained much experience with tobacco cultivation areas from touring the Rhineland, especially in the Dutch and Flemish regions of the Empire in 1802, 1809, and 1810.[17] The late Napoleonic period intensified the circulation of tobacco knowledge in the Rhineland. Reformers in the Alsace benefited from reciprocal contacts between Rhenish projects. Officials from Roer, Baden, Bas-Rhin, but also from Dutch and Flemish Départements addressed and observed each other in letters, books, and other writings.[18] Catherine Maurer and Astrid Strack-Adler have edited a volume that shows the lively cross-border exchange culture of experts and scientists in the Rhineland up to the nineteenth century.[19] In the Napoleonic era, reformers like Schwerz were able to build on a tradition of exchange, trade, and similar mentalities in the Rhineland that Lucien Febvre had already analyzed in his classic study.[20] While upstream Rhine trade lost weight in the Napoleonic period and regional trade centers emerged around Düsseldorf and Mannheim, the view on knowledge circulation shows more reciprocal relations along the river.[21]

The volume of Maurer and Strack-Adler suggests that experts in the Rhineland were intertwined by career paths and professional biographies. Schwerz, too, was not only a traveler, but one of many savants born on the former non-French territories of the Rhine that were able to achieve high administrative posts within areas of Napoleonic Europe that had previously not belonged to the Holy Roman Empire, which had been dissolved in 1806.[22] In June 1812, Schwerz had left his position as head of Coblence's tree nursery to improve tobacco in the Alsace, as he had probably better career prospects attached to the role of "Inspecteur" on the other Rhine side. Neither Napoleonic politics nor administrative tobacco reform were conceivable without the "identification of those individual members of the local elites," whether old or new, "who were judged to possess particular qualities,"[23] expertise to improve local tobacco cultivation in this case. Reformist

bureaucracies had generally intensified cooperation with savants from learned societies around 1800.[24]

Schwerz' position in Strasbourg testifies to the importance of the monopoly administration's resources – specially established posts, infrastructure, and funding –
for the study of tobacco. Established in 1810–1811, the state tobacco monopoly
was organized by the Droits Réunis, a subsector of the Financial Ministry, controlling the cultivation, processing, trade, and purchase of tobacco. The official
monopoly encompassed cultivation areas in parts of Old-France and the newly
occupied Départements alongside the Rhine. The so-called Régie des tabacs was
meant to control cultivation to consolidate Napoleon's finances, ruined by years of
expensive warfare.

Within the borders of the formal French Empire, the Parisian Droits Réunis' obtrusive economic policy was a major catalyst for the mutual recognition of Rhenish
experts. Départements cultivating tobacco for the state were expected to sell raw
tobacco of American quality to the agents of the French Régie. Such visions differed from the eighteenth-century understanding of Rhenish tobacco as an addition,
not a replacement, of resources from the Americas.[25] The resulting pressure on officials to safeguard departmental tobacco cultures in the tobacco regime was raised
by the limit of only ten Départements that were licensed to produce for the Régie.
After 1810–1811, this new pressure situation led to a marked increase in cooperation between agricultural tobacco experts from the circles of learned societies and
the administrators of the imperial Départements.

Paris' economic policy also affected improvement cultures in the Confederation of Rhine, which was not formally part of the monopoly sphere, but shaped by
France's economic influence. "Facing the verbal announcement that a tobacco Régie had been introduced in France and all exports of tobacco leaves were banned,"
officials in the Grand Duchy of Baden had already discussed in January 1811 how
to "take precautions" to ensure that "Baden's lands would not suffer any shortage
of the tobacco leaves they needed themselves."[26] Officially, Baden had been cut off
from the former raw tobacco trade with areas, such as the northern Rhineland, that
were now part of the formal French Empire.[27] Agricultural experts and state servants, however, used debates about possible "scarcity" of colonial goods to evoke
technical-scientific possibilities of import substitution programs. They drama-
tized fears to improve their own reputation, although merchants *de facto* opposed
state trade restrictions and economic warfare by expanding tobacco smuggling. In
Baden, experts pleaded for an expansion of agricultural tobacco improvement initiatives in the Grand Duchy that would be beneficial for the state and its population.

The continental system, i.e., the blockade of British trade to the European continent between 1806 and 1813, had already pushed scientific investigation to avoid
shortages.[28] In Roer, savants from the Société d'Émulation et de l'Agriculture à
Cleves, formed in 1806, immediately began to collaborate with the administration.[29] In the Grand Duchy of Baden, officials relied on various savants, especially
from the traditional tobacco areas around Heidelberg, which had become part of
the enlarged state of Baden after 1806.[30] The pressure situation built up after 1806
and intensified in 1810–1811 with the establishment of the French state tobacco

monopoly jolting many reform-oriented institutions in the Rhineland, such as Strasbourg's prefecture, out of their indifference to tobacco cultivation.

During the French Revolution, still, local farmers had often been the only persons that travelers could ask about Alsatian tobacco cultivation.[31] Not all Rhenish localities can be compared to the Electoral Palatinate, where Theodor II's tobacco monopoly was known for its particular agronomic dynamism since the 1780s.[32] Despite the general rise of government projects and learned writings aimed at improving tobacco cultivation in the late eighteenth century, cooperation between agronomists and administrators to improve tobacco was by no means universal. In the Rhine region, an extensive intensification of tobacco research only started in the Napoleonic period.

Rhenish Botany

With tobacco research's take-off after 1810–1811, adaptations of classical Linnaean eighteenth-century binominal taxonomy of tobacco plants became central elements of reform projects. What was specific to the Rhine region in this research was that botanical analyses were not combined with agricultural chemical approaches, as was quite common elsewhere in the Napoleonic Empire.

Tobacco botanists also set themselves apart from Linnaeus's. His binominal system distinguished between the species' genus and specification – *Nicotiana tabacum* or *Nicotiana rustica* –, while alternative naming was merely tolerated, but not actually regarded as an object of science.[33] In contrast to Linnaeus, Rhenish savants established correlations between the Linnaean system and categories of the Atlantic tobacco trade. Schwerz correlated names such as "Le Grand Asiatique," "Grand Tabac de Virginie," "Varinas," "Tabac d'Amersfort," "Le Brésil," or "Porto Rico" to the respective Linnaean category and added recommendations for the specific use of the variety for manufactural processing.[34]

These correlations were informed by merchant- and farmer-knowledge about tobacco quality (I discuss the importance of practical knowledge for tobacco botany in more detail below). For farmers and merchants, trade terminology for tobacco varieties indicated information on the quality of the taste, proneness for disease, or biotic "degeneration," a vividly discussed issue at the time.[35] Experts hoped that correlations would guarantee that farmers, who were not expected to learn Linnaeus's terminology, were able to keep track of scholarly research. Rapid development of new product names during the period, however, complicated Rhenish attempts to establish such correlations.[36]

Although Rhenish tobacco expertise was botanical, researchers critically addressed the label "botany." "The botanists," as the Strasbourg-born tobacco manufacturer Sinsheim put it, "make their distinctions between species and varieties only with regard to the external forms of the plants and their parts [...]. For the botanist, therefore, the strongest Havana tobacco plant is no different from the weakest North German tobacco plant."[37] For Sinsheim, "botany" tended to leave out applied, agricultural, and consumer-oriented facets, which were important to him and his Rhenish fellows. Contemporaries often poked fun at the purely "theoretical" universal botany.[38] This rhetoric was in line with late eighteenth-century cameralist authors that had begun to conceptualize an "economic botany," a new discipline concerned with the "final purposes" of plants, additional to classical taxonomy.[39]

Outside the Rhineland, American independence encouraged the creation of a chemically oriented tobacco improvement that expanded in parts of Europe during the Napoleonic era. In Berlin and Paris, Sigismund Friedrich Hermbstädt, associate professor of technology and technological chemistry at the University of Berlin since 1810, and Alexis-Antoine Cadet de Vaux, appointed general inspector of the Napoleonic state tobacco manufactures and renowned agricultural publicist, were central representatives of nascent circles of industrial and agricultural chemists, most of them academically trained.[40]

These reformers made strong claims about the effectiveness of Lavoisier's new chemical theories in agriculture: chemistry, as Cadet de Vaux stated, reduced obstacles to the enhancement of tobacco plants. It was believed that the "deficiencies of nature," i.e., the colder temperatures in Europe, could be compensated for by the "physical sciences," chemical methods in particular.[41] Elements such as ammonia, a product of nitrogen, which was mainly found in urban latrine excrements, were seen as keys to successful plant growth.[42] On the threshold of the nineteenth century, the contours of an agrochemical discipline communicating via a specialized, internationally connected publication culture emerged which also rubbed off on tobacco research.[43] This chemical tobacco research appeared primarily locally, in scientific centers such as Berlin and Paris.

Savants in other parts of the Empire, such as the Rhineland, often remained rather distanced from such innovations. Strasbourg's tobacco inspector Schwerz was "very fond of art and scientific empiricism" in agriculture, but saw little value in the application of "actual science" (which included agricultural chemistry in his view).[44] This attitude reflected the widespread rejection of "applied science" in reform circles.[45] In Alsace, optimists toward chemical approaches remained a minority.[46] As others, Schwerz was an advocate of inductive methods.[47] Tobacco experts in the Rhineland referred to rather empirical explanations for the utility of fertilizers or to the vitalist humus theory, while savants in Paris and Berlin mobilized the natural sciences.

Reluctance toward chemistry went hand in hand with a thorough integration of farmers into Rhenish reform projects. Reformers of Strasbourg's Société des Sciences, Agriculture et Arts du Bas-Rhin, which had been formed in 1799,[48] based their tobacco cultivation research on "various pieces of information" from the "best farmers," who were interviewed by scholars.[49] In Roer, tobacco inspectors consulted the farmers Guillaume van Laak and Jean Voss for the drafting of a handbook.[50] References to the value of farmers' knowledge indicate a shift in the perception of learned agrarian reformers and state officials around 1800: the figure of an "enlightened" or at least teachable farmer gradually replaced the one-sided stereotype of the inaccessible "Sudelwirt" ("scraw peasant"). The rural population came to be seen in a more positive light.[51] Regina Dauser and others have shown how agricultural reform projects around 1800 depended on the active involvement from "below," on the support of farmers.[52] In this regard, the increasing exchanges among tobacco experts in the Napoleonic Rhineland can also be understood as a form of mobilization of the knowledge that farmers had shared with Rhenish officials in petitions and letters.

Agricultural chemists such as Cadet de Vaux or Hermbstädt remained more aloof to farmers' practical spheres of knowledge. This is evidence for early nineteenth-century "boundary work," a rhetoric of distance toward practically oriented persons.[53] It differed from the culture of tobacco research in the Rhineland, where hierarchies were motivated by a different logic: here, social etiquette prevented farmers from being accepted directly as members of learned agricultural societies whose membership was recruited from middle-class and aristocratic circles, while their knowledge was incorporated in the regional knowledge culture.[54] Boundary rhetoric stabilized the developing field of agricultural chemistry. In this sense, botanical Rhenish tobacco research was clearly more strongly oriented toward the practical sphere of knowledge of farmers. Scientific regionality manifested in a Rhenish geography of improvement with participants that shared a particular botanic style, differing from other contemporary pursuits, chemical research, in particular.

Patronage Relations

Patronage relationships, common within the Napoleonic power apparatus, promoted the botanic and disadvantaged the chemical nature of Rhenish tobacco research. Without personal ties to Bas-Rhin's prefect, Adrien de Lezay-Marnésia, whom he called his "friend" and "benefactor,"[55] Strasbourg's tobacco inspector Johann Nepomuk Schwerz would probably not have been considered for the office of tobacco inspector in Bas-Rhin. Both men knew each other from their former cooperation in the Département of Rhin-et-Moselle where Lezay-Marnésia had been prefect between 1806 and 1810. There, he had been Schwerz's superior while Schwerz worked as managing director for the Département's tree nursery in Koblenz. In Alsace, where he had been prefect since 1810, Lezay-Marnésia entertained similar patronage relations with important families of the region.[56] It was due to patronage that the Alsatian tobacco research culture was led by a skeptic of agricultural chemistry who, if he intended to use natural sciences for reform purposes, gave preference to botany.

As a basis for the appointment of state experts in the Rhineland, patronage also had exclusionary effects. Inspecting Strasbourg's state tobacco manufacture during a visiting tour to the Empire, Alexis-Antoine Cadet de Vaux, chemist and appointed general inspector of the imperial state tobacco manufactures, also conveyed his knowledge on agronomy and chemical improvement to the local prefecture.[57] There is evidence that Rhenish savants would have appreciated Cadet de Vaux's engagement in the state-run reform projects as feedback on Cadet de Vaux's tobacco literature in the Empire's print public was positive.[58]

During the inspection-visit, however, rumors begun to spread in the Alsace that Cadet de Vaux planned to capitalize on a secret sauce recipe of Strasbourg's tobacco manufacture after returning from his inspection tour to Paris.[59] Alsatian elites perceived this recipe as an important factor for the region's competitiveness in the snuff tobacco market. The sauce was seen as the hallmark of the former eighteenth-century private tobacco business.[60] Cadet de Vaux failed to convince the prefect of his agro-chemical agenda, because his leadership of Strasbourg's tobacco improvement culture would not only have jeopardized the prefecture's clientele relationship with the botanist Schwerz, but also with regional notables that worked in Strasbourg's public tobacco administration. Among these regional notables were previously expropriated tobacco manufacturers, as the director of the manufacture Ange-Marie Francois Gaétan Marocco, who now tried to rescue and safeguard their production secrets.[61] Cadet de Vaux had no lobby among these elites. Consequently, he was unable to put into effect a formal cooperation and Parisian agricultural chemistry as the leading paradigm of Alsatian tobacco research.

Missing patronage relationships also limited the intensity with which reformers connected on regional scale. During the spring and summer of 1812, Rhenish members of the Société d'Émulation et de l'Agriculture à Cleves criticized a French-born monopoly official by the name Gruet, "Contrôleur principal" of the Département de la Roer, one of the northern Départements founded after French occupation in the 1790s, for excluding them from the production of a tobacco cultivation guidebook.[62] Confronted with these complaints, Gruet felt compelled to take offensive action against its members and their feeling of "bitterness."[63] While the Société d'Émulation cooperated with the Département's tobacco administration in many other instances, its representatives were victims of the disadvantaged position of not being French-born citizens. Native Frenchmen were often structurally favored over foreign-born individuals.[64]

These reactions highlight prejudices about the traditional elites: aggressive perceptions of the native Rhinelanders as unenlightened were not uncommon among French-born administrators and savants that understood reform as France's assigned "civilizing mission."[65] This resulted in the confusing situation that native Rhenish elites often did not get a chance to form patronage relationships with the higher administration, which was largely staffed by native Frenchmen. In the specific case of the conflict over Gruet, not Rhenish elites, but farmers were preferred as informants and sources. Gruet would generally not have considered patronage with socially lower farmers – which, of course, continued to be suspected of needing to be "civilized." Patronage with socially equal scholarly cooperation partners, as the Société d'Émulation, was also complicated because of the exclusive effects of French cultural prejudices.

Disintegration in regional tobacco improvement was finally also motivated by a feeling of marginalization and patronage-unwillingness that savants expressed when they faced the new French-modeled state institutions. In January 1811, the Constance Lake District Directorate (Seekreisdirektion) of the Grand Duchy of Baden asked Karlsruhe's court botanist, Carl Christian Gmelin, a well-known figure of the Agricultural Enlightenment in the southern German states, to compose a manual for regional tobacco cultivation. The botanist saw it as a "rare affront [...] to be called upon by [...] [a] district director" (Kreisdirektor) for possible cooperation.[66] As director of the Margrave's Natural History Cabinet and the botanical

gardens in Karlsruhe, Gmelin was not financially or career-wise dependent on tobacco research and probably saw in it only a secondary occupation – quite the
opposite of Schwerz. Gmelin rejected the demand of the Lake District Directorate,
because he was irritated by the administrative competences in agriculture which
had been assigned to Baden's district directorates after the organizational edict of
1809, which was oriented to the competences of the French prefectures.[67] Being
socialized in the administrative system of the prerevolutionary period, savants such
as Gmelin expressed deep concern with the creation of economically proactive
Napoleonic bureaucracies apart from the capitals' central administration and thus
also ruled out patronage relationships.

Parisian Passivity

Patronage was a crucial force of scientific regionalization in the Empire, favoring
a plural research landscape. But Rhenish tobacco knowledge was further strengthened by the fact that the Parisian central administration of the state tobacco monopoly hardly promoted an empire-wide, chemically motivated tobacco improvement
policy. The Droits Réunis' central Parisian tobacco office was a place that distributed knowledge resources to the imperial peripheries, not a hub for centralized
policymaking. Officials from Paris provided financial and logistical support for
tobacco improvement projects which were otherwise shaped by regional networks
and exchange between the Empire's Départements and affiliated states.[68] Upon
request by departmental authorities, Parisian officials forwarded seeds and documents on tobacco plants from other territories, encompassing Dutch and Belgian
regions.[69]

The Parisian Droits Réunis supported most initiatives but held only a few ones
in high regard. Antoine Français de Nantes, director-general of the Droits Réunis since 1804 and one of the most important actors in the central administration
of the monopoly, particularly esteemed Alsatian tobacco projects. For de Nantes,
Bas-Rhin's departmental prefect Adrien de Lezay-Marnésia was an "enlightened
official" whose "measures to increase imperial revenues" were essential for the
state's fiscal prosperity.[70] Lezay-Marnésia's *Instruction* was officially disseminated
to the enlightened public of the Empire.[71] Also journals and newspapers singled out
Lezay-Marnésia's merits for agricultural development in the Département Rhin-et-
Moselle.[72] While other Napoleonic officials enjoyed support from Paris, they were
far from being treated in comparable ways. Regional initiatives in the imperial
Rhineland were supported by a Parisian policy that was economically obtrusive,
yet indifferent toward centralized improvement designs.

Employment continuities in the state's financial sector were important in this
regard. The Droits Réunis were staffed by several former members of the *fermes
générales*. Existing until the French Revolution, the *fermes* were an institution to
which the French king had leased the right to organize the tobacco industry of the
monarchy. Among these former "*fermiers*" were Français de Nantes, who had organized a directorate of the five *fermes* in the city of Nantes from 1778 to 1790,[73] or
François-Nicolas Mollien, who had joined the *fermes* in the same year.[74] Both were

part of a small minority of the 40 to 60 *fermiers* who survived persecution during the revolutionary period, were granted amnesty by Napoleon, and came to occupy important position in his higher financial administration.[75]

The attitude of the former monarchical state monopoly staff transformed during the Napoleonic era. Post-Napoleonic contemporaries noticed that the passivity of Parisian civil servants was like France's prerevolutionary *fermes*.[76] This is only partially true, because before the French Revolution there were no approaches at all to promote tobacco cultivation in the kingdom through improvement measures. Since the 1720s, a ban on tobacco cultivation that even encompassed pharmacy and monastery gardens had been in place in French European territories to prevent smuggle, one of the *fermiers'* central fears.[77] Wars, the Continental Blockade, and autarchy fantasies moved the former *fermiers* away from their fundamental skepticism about tobacco cultivation. As a relic of the prerevolutionary period, however, an indifference to improving tobacco cultivation in European France remained in the Parisian Droits Réunis which kept them from designing and implementing empire-wide, centralized projects. Oriented toward the old system of the *fermes gé-nérales*, the Parisian Droits Réunis only focused on modernizing the manufactures, which became public property with the establishment of the Napoleonic tobacco monopoly.[78]

For some of the many tobacco cultivation experts that gained public ground between the French Revolution and the "reintroduction" of the state tobacco monopoly in 1810–1811, this approach was not enough. Cadet de Vaux demanded a more centralized improvement policy for tobacco cultivation in the Empire: "In this epoch of vast conquests, which extended its dominance and ensured the service of all climates, France is able, more than ever before, to free itself from the import of foreign tobaccos. Not only can its own tobacco consumption be guaranteed, but tobacco cultivated in the soil of the Empire can also become an important object of cultivation, industry, and trade with other peoples of Europe."[79]

Cadet de Vaux imagined a space of action that oscillated between a national and an imperial understanding of France: the Rhine formed the French border in the style of revolutionary rhetoric,[80] yet a "France" reaching the distant territories controlled by Napoleon was also discernible. Cadet de Vaux's publication shows how French-speaking agronomic tobacco researchers tried to make sense of the new possibilities they assigned to the imperial monopoly system.

Experts such as Cadet de Vaux might have successfully realized their plans if the French Ministry of the Interior, and not the Financial Ministry, to which the Droits Réunis were attached, had become responsible for the tobacco sector. Already before the creation of the state tobacco monopoly, the Ministry of Interior had shown a keen interest in the crop. In the slipstream of Napoleons's conquests in North Africa between 1798 and 1801, Jean-Antoine Chaptal, head of the Ministry of the Interior, sent an agricultural mission to Egypt to foster tobacco cultivation in the Nile region, among other economic projects.[81] The strict administrative division of the postrevolutionary ministries, ensuring the autonomous work of state institutions, prevented these dynamics from spilling over to the Droits Réunis. Central passivity remained, leaving agency for state reformers in the Départements.

Preferring to focus on those projects in which Parisian scholars and administrative authorities pursued centralized projects for the agricultural transformation of the Empire,[82] historians have too long ignored the autonomous activities in the peripheries. The example of tobacco cultivation shows that regional initiatives, the networks, and knowledge cultures based on them were often promoted by a wait-and-see, passive attitude from the centers. Regional tobacco researched flourished due to support of a passively acting imperial administration center. While a number of tobacco researchers publicly promoted a centralized culture of reform, the continuities in personnel between the tobacco monopoly of the Ancient Regime and the Napoleonic era prevented that.

Nationalized Networks

The so-called Wars of Liberation since 1813 and the later dissolution of the Napoleonic Empire finally fostered national loyalty and changed the conditions under which Rhenish reformers debated regionally. Official regional cooperation, whether in terms of employment or formal institutional alliances, became increasingly difficult in view of the national commitments demanded. Although this did not tackle the coherence in the botanical reform culture in the Rhineland, which continued to exist after 1815, it fostered a more informal regional exchange between Rhenish experts.

In 1814, Schwerz decided to leave Strasbourg's tobacco reform project to move back to his German "fatherland" and join comparable activities on tobacco improvement and Agricultural Enlightenment. In his *Pfälzer Reise* ("Voyages in the Palatinate") published in the same year, the Koblenz-born savant described the special feeling of treading on "genuinely German soil" again in Offenbach, Grand Duchy of Hesse. Only a few months after his departure from Strasbourg, the work in Alsace seemed little more than an embarrassing, unpatriotic episode in his career as a tobacco expert: "For a man who loves his people and fatherland and is not a sold hireling and indulger of strangers," he wrote, life outside Germany "is not indifferent and, if he has been away for a long time, makes no small impression on his heart."[83]

Schwerz's motivation for leaving was certainly also due to the fact that Lezay-Marnésia, his patron and benefactor, had died in the same year and relations with the regional elites thus required complicated renegotiation. The consequences of nationalization, however, were unmistakable in the region. Movements of agricultural experts were reinterpreted as unpatriotic misconduct when Napoleonic rule began to fade away.[84] In France as in the German states, the Napoleonic wars strengthened the reciprocal perception as a national "arch-enemy."[85] The establishment of the Napoleonic administration and the political dominance of France had prepared ground for national sentiments. While reformers from the former Holy Roman Empire were still able to work officially in Strasbourg's tobacco administration in 1812, this became increasingly difficult with the national reorientation on both sides of the Rhine.

The late Napoleonic era represented an acceleration of a long-standing trend in this regard. National orientations were by no means new in the years after 1813, but had already been known to tobacco researchers, merchants, and manufacturers in the late eighteenth century. When plans for a future tobacco monopoly of the French state became known in the 1790s, a politicized discussion between traders along the Rhine ensued. The traders resisted the increasing tax burden on tobacco and the threat of monopolization. Cologne's manufacturers had, in a national interpretation, counted their Strasbourg counterparts among those of "old France," emphasizing their shared interests in contrast to those of the "Rhinelanders."[86] During the earlier eighteenth century, areas such as Alsace were still understood as transitional zones, or frontiers, between different cultural (language) spaces in Europe. They became increasingly perceived as a sharp border between two clearly separated national areas in the nineteenth century – Schwerz's reference to Offenbach as the border of Germany was less sharp than the common contemporary image.[87]

In the years immediately before and after the Congress of Vienna, national sentiments still overlapped with memories of the lucrative opportunities that the imperial state tobacco reform had offered to experts from the former Holy Roman Empire. On 14 March 1815, only a few months after his departure from Strasbourg, Schwerz tried to reestablish contact with the tobacco administration in Strasbourg and reactivate his former employment in the Département. Schwerz offered his services and stated his wish to be reinstated in the administration of the Alsatian tobacco monopoly "in a manner befitting his knowledge."[88]

His application testifies to his hope for a revived imperial integration of tobacco research in the Rhine region, which for contemporaries briefly gained new urgency in the face of Napoleon's attempts to establish a new empire. It should be noted that Schwerz's candidacy came exactly 14 days after the beginning of Napoleon's Hundred Days. Against the backdrop of the successful cooperation between Schwerz and Alsatian officials during Napoleon's first reign, Schwerz's reinstatement seemed not only realistic but also to have a long-term prospect. Although Napoleon's defeat at Waterloo on 22 June 1815 ultimately prevented Schwerz's reintegration into the Alsatian tobacco administration, the contacts indicate a continuing, reciprocal interest on the part of scholars and administrators, even across the political caesura. There was a certain form of nostalgia for the reforms of the Empire, which was expressed in the southern Rhineland regions above all in the continuing admiration and mystification of Strasbourg's prefect Lezay-Marnésia.[89]

No formal cooperation existed between tobacco experts in the region after the Napoleonic Empire collapsed. Nonetheless, exchange of knowledge about tobacco continued in the Rhineland, although the discussion centered more and more on the Upper Rhine valley. Chaired by Schwerz's successor as tobacco inspector in Strasbourg, J. B. N. L. Husson, the Société des Sciences, Agriculture et Art du Bas-Rhin re-enacted experiments with tobacco conducted by researchers from the Landwirthschaftliche Verein für das Großherzogtum Baden in the 1820s.[90] Experts continued to build on relationships in the Rhine region after the Napoleonic era, although the tobacco-growing areas of the Lower Rhine progressively disappeared from the scene in the course of the nineteenth century. Exchange of tobacco knowledge between Rhenish regions continued well into the last third of the nineteenth century, but the political integration of the Napoleonic Empire had faded away from one day to the other.

Conclusion

Knowing Empires were shaped by spatial organization of knowledge circulation, in regard to its nationalization, as one might think with regard to recent studies, but also on the regionalization. As the chapter shows with the example of the Napoleonic imperial conquests in Europe during the early nineteenth century, mechanisms of Napoleonic statehood played a major role in the development of a regional botanical tobacco research culture in Alsace, the southern Rhine regions of Baden and the Palatinate, as well as the North Rhine tobacco-growing areas around Kleve and Cologne, in which experts and reformers were closely connected. The French tobacco monopoly administration that emerged in 1810–1811 and the reform administrations installed in the affiliated and dominated German-speaking areas, which were strongly influenced by France, were central nodes of this regional exchange. Patronage relations, central to the emergence of the Napoleonic power apparatus, prevented Alsatian tobacco research, which was a model for other Rhenish localities, from being shaped by representatives of the agricultural chemistry that had been developing in some places in Europe since the late eighteenth century. Botany's strong influence on the Rhenish culture of knowledge was further supported by the lack of centralized, empire-wide measures taken by the Parisian administration of the tobacco monopoly. On this basis, a regional culture of knowledge developed in the Rhineland, which was not significantly affected by the intensified nationalization processes of the later Napoleonic period. Showing similar thought patterns and connections in the Rhineland, the chapter analyzed how state mechanisms of the forming Napoleonic empire helped to form a regional space of circulating tobacco. Historians of science and knowledge may take this as a starting point to shed light on the geographic-spatial aspects of knowledge circulation in empires, in Europe as in the rest of the world, abandoning the often rather one-sided focus on the nationalization of knowledge circulation that has long characterized research.

Notes

1 This argument is a development of material contained in Chapters 2 and 3 of Alexander van Wickeren, *Wissensräume im Wandel. Eine Geschichte der deutsch-französischen Tabakforschung (1780–1870)* (Cologne, Weimar and Vienna: Böhlau, 2020).
2 Peter M. Jones, *Agricultural Enlightenment. Knowledge, Technology, and Nature, 1750–1840* (Oxford: Oxford University Press, 2015).
3 [Adrien de Lezay-Marnésia], *Instruction für die Tabakspflanzer, zum Gebrauche des Niederrheinischen Departements,* [Strasbourg], 12 September 1811. Archives départementales du Bas-Rhin (ADBR), 15/M/402, unpaginated; Gruet, *Handbuch des Tabaks-Bauers, zum Gebrauch des Klevischen und Kölnischen Bezirks, im Roer-Departement* (Aachen, 1812); in the same year, in the Grand Duchy of Baden, one of the vassal states of the Confederation of the Rhine founded by Napoleon in 1806, the mayor of Constance, Anton Burkhart, published a similar manual: Entry of the Financial Ministry, Karlsruhe, 5 October 1811, Generallandesarchiv Karlsruhe (GLA), 236/6072, Fol. 25.

4   Kapil Raj, "Beyond postcolonialism... and postpositivism: Circulation and the global history of science," *Isis* 104 (2013): 337–347.
5   On the particular relation between state, knowledge, circulation, and space, see: Lothar Schilling and Jakob Vogel, eds., *Transnational Cultures of Expertise. Circulating State-Related Knowledge in the 18th and 19th Centuries* (Berlin: De Gruyter, 2019).
6   Pascal Schillings and Alexander van Wickeren, "Towards a Material and Spatial History of Knowledge Production: An Introduction," *Historical Social Research* 40 (2015): 203–218; Regina Dauser and Lothar Schilling, "Einleitung. Raumbezüge staatsrelevanten Wissens," in: discussions, 7 (2012) – Grenzen und Kontaktzonen, http://www.perspectivia.net/content/publikationen/discussions/7-2012/dauser-schilling_einleitung, [consulted: 19 September 2019]; Diarmid A. Finnegan, "The Spatial Turn: Geographical Approaches in the History of Science," *Journal of the History of Biology* 41(2008): 369–388; David N. Livingstone, *Putting Science in its Place. Geographies of Scientific Knowledge* (Chicago, IL: University of Chicago Press, 2003); Mitchel G. Ash, "Räume des Wissens – was und wo sind sie? Einleitung in das Thema," *Berichte zur Wissenschaftsgeschichte* 23 (2000): 235–242; Crosbie Smith and Jon Agar, eds., *Making Space for Science: Territorial Themes in the Shaping of Knowledge* (Basingstoke: Palgrave Macmillan, 1998); David N. Livingstone, "The Spaces of Knowledge. Contributions Towards a Historical Geography of Science," *Environment and Planning* 30 (1995): 5–34; Adi Ophir and Steven Shapin, "The Place of Knowledge: A Methodological Survey," *Science in Context* 4 (1991): 3–21.
7   Alexander Wickeren, "Territorializing Atlantic Knowledge: The French State Tobacco Monopoly and the Globalization of the Havana Cigar around Mid-19th Century," in Lothar Schilling and Jakob Vogel, eds., *Transnational Cultures of Expertise*, 181–196; Marianne Klemun, "Geognosie versus Geologie. Nationale Denkstile und kulturelle Praktiken bezüglich Raum und Zeit im Widerstreit," *Berichte zur Wissenschaftsgeschichte* 38 (2015), 227–242; Mitchell G. Ash and Jan Surman, "The Nationalization of Scientific Knowledge in Nineteenth-Century Central Europe. An Introduction," Mitchell G. Ash, ed, *The Nationalization of Scientific Knowledge in the Habsburg Empire, 1848–1918,* (Basingstoke: Palgrave Macmillan, 2012), 1–29; Mario Ackermann, *Wissenschaft und nationaler Gedanke im 18. und frühen 19. Jahrhundert. Eine Studie zum Nationalismus am Beispiel des Gedankenguts der deutschen Forscher Johann Beckmann und Johann Friedrich Ludwig Hausmann im Kontakt mit schwedischen Gelehrten 1763 bis 1815* (Münster: Lit Verlag, 2009); Ingrid Merchiers, *Cultural Nationalism in the South Slav Habsburg Lands in the Nineteenth Century. The Scholarly Network of Jernej Kopitar (1780–1844)* (München: Sagner, 2007), Ralph Jessen and Jakob Vogel, "Einleitung. Die Naturwissenschaften und die Nation. Perspektiven einer Wechselbeziehung in der europäischen Geschichte," Ralph Jessen and Jakob Vogel, eds., *Wissenschaft und Nation in der europäischen Geschichte* (Frankfurt Main: Campus Verlag, 2002), 7–37; Kai Torsten Kanz, *Nationalismus und internationale Zusammenarbeit in den Naturwissenschaften. Die deutsch-französischen Wissenschaftsbeziehungen zwischen Revolution und Restauration 1789–1832* (Stuttgart: Steiner Verlag, 1997).
8   See, on a more general level: Charles W. J. Withers, *Placing the Enlightenment: Thinking Geographically About the Age of Reason* (Chicago, IL: University of Chicago Press, 2007), 34; Livingstone, *Putting Science in its Place,* 87–88; for a case study, see: Wilhelm Kreutz, *Aufklärung in der Kurpfalz. Beiträge zu Institutionen, Sozietäten und Personen* (Ubstadt-Weiher: Verlag Regionalkultur, 2008), 35–70.
9   Simon Naylor, *Regionalizing Science. Placing Knowledges in Victorian England* (London: Pickering & Chatto, 2010).
10 van Wickeren, *Wissensräume im Wandel.*
11 Frederick Albritton Jonsson, "Scottish Tobacco and Rhubarb: The Natural Order of Civil Cameralism in the Scottish Enlightenment," *Eighteenth-Century Studies* 49 (2016): 129–147; Regina Dauser, "Das Wissen der Herrschaft. Wissensgenerierung und Reformen der praktischen Aufklärung in der zweiten Hälfte des 18. Jahrhunderts,"

in Mark Häberlein, Stefan Paulus, and Gregor Weber, eds., *Geschichte(n) des Wissens. Festschrift für Wolfgang E. J. Weber zum 65. Geburtstag* (Augsburg: Wißner, 2015), 619–633, here 627–631; Curtis Carroll Davis, "'A National Property:' Richard Claiborne's Tobacco Treatise for Poland," *The William and Mary Quarterly* 21 (1964): 93–117.

12 On the eve of the French Revolution, roughly 90% of tobacco imported to France came from North America – not from the French oversea possessions: Marc Vigié and Muriel Vigié, *L'herbe à Nicot. Amateurs de tabac, fermiers généraux et contrebandiers sous l'Ancien Régime* (Paris: Fayard, 1989), 153–180 and 187. If one compares the number of German-language tobacco works published in the period that goes from the early seventeenth century to 1776 to the period between 1776 and 1800, one notices a distinct growth in this kind literature in the late eighteenth century. While in the first phase only 26 books were published, 40 writings on tobacco cultivation appeared in the much shorter period between 1776 and 1800: Rainer Immensack, *Bibliographie als Geschichte der deutschsprachigen Tabakliteratur von 1579-1995* (Braunschweig: Brandes, 1996), 1–8, 39, and 44–46.

13 See the conference: Rethinking tobacco history: Commodities, empire and agency in global perspective, 1780–1960. In: H-Soz-Kult, 02.09.2020, www.hsozkult.de/event/id/event-93187 (consulted on 24 May 2022).

14 Jordan Goodman, *Tobacco in History: The Cultures of Dependence* (London: Routledge, 1993), 193–215.

15 Günther Franz, *Johann Nepomuk Schwerz. Gedächtnisrede anlässlich der 200. Wiederkehr seines Geburtstages bei der Jahresfeier der landwirtschaftlichen Hochschule am 20. November 1959* (Stuttgart: Ulmer, 1960), 10.

16 Schwerz, "Note sur les tabacs cultivés au jardin économique" (without date or place), ADBR, P/259, unpaginated; idem, "Ordre à suivre tant dans la conduite des jardins destinés 1. à l'éducation du plant & des porte-graines de tabac, que 2. dans celle des champs destinés aux expériences de culture, & à celles des 3. differents modes de dessication" (without date or place), ADBR, P/259, unpaginated.

17 Ibid., 8–9.

18 Letter, Receveur général to Lezay-Marnésia, Bruges, 16 March 1812, ADBR, 11/M/196, unpaginated; Gruet to Lezay-Marnésia, Cleve, 7 May 1812, ADBR, 11/M/194, unpaginated; Gruet to Lezay-Marnésia, Cleve, 10 January 1813, ADBR, 11/M/197, unpaginated.

19 Catherine Maurer and Astrid Strack-Adler, eds., *L'espace rhénan, pôle de savoirs* (Strasbourg: Presses universitaires de Strasbourg, 2013).

20 Lucien Febvre, *Der Rhein und seine Geschichte* (Frankfurt/Main and New York: Campus, 1994).

21 Robert Mark Spaulding, "Rhine River Commerce and the Continental System," in Katherine B. Aaslestad and Johan Joor Basingstoke, eds., *Revisiting Napoleon's Continental System: Local, Regional and European Experiences* (Basingstoke: Palgrave Macmillan, 2015), 114–134.

22 Gabriele B. Clemens, "Verwaltungseliten und die napoleonische Amalgampolitik in den linksrheinischen Departements," in Willi Jung, ed., *Napoléon Bonaparte oder der entfesselte Prometheus* (Bonn: V & R Unipress, 2015), 67–92.

23 Woolf, *Napoleon's Integration*, 107; see also Michael Rowe, *From Reich to State: The Rhineland in the Revolutionary Age, 1780–1830* (Cambridge: Cambridge University Press, 2003), 114: "The most important characteristic of the Napoleonic government was less its centralisation and more its dependence upon local elites."

24 Stefan Brakensiek, "Das Feld der Agrarreformen um 1800," in Eric J. Engstrom, Volker Hess, and Ulrike Thoms, eds., *Figurationen des Experten. Ambivalenzen der wissenschaftlichen Expertise im ausgehenden 18. und frühen 19. Jahrhundert* (Frankfurt/Main: Lang, 2005), 101–122, here 119.

25 Lezay-Marnésia, *Instruction*, 19.

26 Entry of Baden's Ministry of Interior, Landes Oeconomie Departement, Karlsruhe, 19 January 1811, GLA, 236/16731, Fol. 5–6; with reference to the "lack of tobacco" see also: "Die Tabaks- und Hopfenkultur betreffend. Auf mündlichen Vortrag, daß der zur Emporbringung des Tabakes und Hopfenbaues im Seekreis anher berufene Joseph Brucker von Brühl Amts Schwetzingen dahier angelangt sey, wird vermöge Beschlußes den sämmtlichen Aemtern und Gefällverwaltungen des Kreises eröfnet," Konstanz, 26 April 1811, GLA, 236/6072, Fol. 17.

27 Heinrich Hassinger, *Der oberbadische Tabakbau und seine wirtschaftliche Bedeutung* (Karlsruhe: G. Braunsche, 1912), 7–9, points out that raw tobacco exported from Alsace to Baden could no longer be supplied. Michael Rowe, on the other hand, believes that the continental system had a particularly beneficial effect on west-east trade within the Napoleonic Empire. This may be true for many products, but not for tobacco after the creation of the French state tobacco monopoly in 1810–11: Rowe, *From Reich to State*, 199.

28 van Wickeren, *Wissensräume im Wandel*, 57–58.

29 Gustav Mücke, *Die geschichtliche Stellung des Arrondissements und seines Verwalters zur Zeit der napoleonischen Herrschaft. Dargestellt am Beispiel des Lebens und Wirken Karl Ludwig von Keverbergs als Unterpräfekt in Cleve* (PhD dissertation: Universität Bonn, 1935), 121–125.

30 See the discussion between Fol. 11 and Fol. 15, Fol. 32 and Fol. 50 in the Fascicel GLA, 236/16731.

31 Anonymous, *Reise eines Engländers durch einen Theil von Elsaß und Nieder-Schwaben. In Briefen verfaßt, und von seinem teutschen Freunde L. U. F. D. B. herausgegeben* (Amsterdam, 1793), 13–14.

32 Dauser, "Das Wissen der Herrschaft."

33 Staffan Müller-Wille, "Ein Anfang ohne Ende. Das Archiv der Naturgeschichte und die Geburt der Biologie," in Richard van Dülmen and Sina Rauschenbach, eds., *Macht des Wissens. Die Entstehung der modernen Wissensgesellschaft* (Cologne: Böhlau, 2004), 587–605, here 600.

34 Schwerz, "Note sur les tabacs cultivés au jardin économique."

35 For merchants, the regional origin and quality of the leaves were closely related. With a view to early modern times and the nineteenth century, see: Barbara Hahn, "Paradox of Precision: Bright Tobacco as Technology Transfer, 1880–1937," *Agricultural History* 82 (2008): 220–234, here 224.

36 van Wickeren, *Wissensräume im Wandel*, chapter 2.

37 M. Sinsheim, *Die Geheimnisse der sämmtlichen Rauch- und Schnupftabaks-Fabrikation* (Frankfurt/Main, 1826), 5–6.

38 Jones, *Agricultural Enlightenment*, 176.

39 Günther Luxbacher, "Die technologische Mobilisierung der Botanik. Konzept und Wirkung der technischen Rohstofflehre und Warenkunde im 19. Jahrhundert," *Technikgeschichte* 68 (2001): 307–333, here 311–312.

40 On Hermbstädt, see Jakob Vogel, *Ein schillerndes Kristall. Eine Wissensgeschichte des Salzes zwischen Früher Neuzeit und Moderne* (Cologne: Vandenhoeck & Ruprecht, 2008), 302–309; on Hermbstädt's interest in chemical tobacco research, see: Sigismund Friedrich Hermbstädt, *Gründliche Anleitung zur Kultur der Tabackspflanzen und der Fabrikation des Rauch- und Schnupftabacks nach agronomischen, technischen und chemischen Grundsätzen* (Berlin, 1822), 106; on Cadet de Vaux, see Spary, *Feeding France,* 38; August Boerner, *Kölner Tabakhandel und Tabakgewerbe. 1628–1910* (Essen: C. D. Baedecker, 1912), 126.

41 Alexis-Antoine Cadet de Vaux, *Traité de la culture du tabac, et de la préparation de sa feuille, réduites à leurs vrais principes* (Paris, 1810), 75.

42 Ibid., 17; on nitrogen and ammonia, see: G. J. Leigh, *The World's Greatest Fix: A History of Nitrogen and Agriculture* (Oxford: Oxford University Press, 2004), 96–98.

43 Peter M. Jones, "Making Chemistry the 'Science' of Agriculture, c. 1760–1840," *History of Science* 54 (2016), 169–194.

44 Franz, *Johann Nepomuk Schwerz*, 8.

45 Jones, *Agricultural Enlightenment*, 180.

46 In Strasbourg's Société des Sciences, Agriculture et Arts du Bas-Rhin, chemical tobacco research only appeared once, in 1809: "Procès-verbaux des séances de la Société d'Agriculture, Sciences et Arts du Département du Bas-Rhin, an 10-1827, séance générale du 13. Avril 1809," ADBR, 63/J/30, 75–76.

47 Ibid., 18.

48 Letter, Sécrétaire générale de la Préfecture de la Département du Bas-Rhin to Société des Sciences, Agriculture et Arts du Département du Bas-Rhin, Strasburg, 17 October 1812, ADBR, 63/J/3, unpaginated.

49 Letter, Société des Sciences, Agriculture et Arts du Département du Bas-Rhin to Lezay-Marnésia, Strasbourg, 7 May 1811, ADBR, 11/M/194, unpaginated.

50 Letter, Gruet to Préfet du Département de la Roer, Cleve, 21 March 1812, LA NRW, Roerdepartement, 02025, Fol. 34–35.

51 Verena Lehmbrock, "Agrarwissen und Volksaufklärung im langen 18. Jahrhundert. Was sehen historische Gewährsleute und was sehen ihre Historiker/innen?," in Martin Mulsow and Frank Rexroth, eds., *Was als wissenschaftlich gelten darf. Praktiken der Grenzziehung in Gelehrtenmilieus der Vormoderne* (Frankfurt/Main: Campus, 2014), 485–514, here 501–503.

52 Dauser, "Das Wissen der Herrschaft."

53 On the rhetorical demarcation of science and non-science in the nineteenth century, see: Thomas Gieryn, "Boundary-Work and the Demarcation of Science from Non-Science. Strains and Interests in Professional Ideologies of Scientists," *American Sociological Review* 48 (1983): 781–795.

54 Marcus Popplow, "Economizing Agricultural Resources in the German Economic Enlightenment," in Ursula Klein and Emma C. Spary, eds., *Materials and Expertise in Early Modern Europe: Between Market and Laboratory* (Chicago, IL: University of Chicago Press, 2010), 261–287, here 276 and 278.

55 Johann Nepomuk Schwerz, *Beschreibung der Landwirtschaft des Nieder-Elsaß* (Berlin, 1816), vii–viii.

56 von Westerholt, *Lezay Marnesia*, 202–214.

57 Although Cadet de Vaux referred to the need to dry tobacco in drying barns and the problem of over-moist tobacco, we can assume that he used these aspects mainly as a door opener to recommend his agricultural-chemical tobacco research as a guideline for the Strasbourg reform culture: Letter, Directeur général de l'Administration des Droits Réunis to Lezay-Marnésia, Paris, 11 February 1812, ADBR, 11/M/196, unpaginated; on Cadet de Vaux's stay in Strasbourg and the meeting, but without mentioning the talks on tobacco cultivation, see: Alexis-Antoine Cadet de Vaux, *Moyens de prévenir le retour des disettes* (Paris, 1812), 152. Although there is no official letter of application, the fact that Cadet de Vaux corresponded with the scholars of the Société des Sciences in Strasbourg before his trip to Alsace suggests he may have applied for membership: Alphonse M. Koch, "La Société libre d'Agriculture et d'Économie intérieur du département du Bas-Rhin du 15 floréal an VIII au 4me jour complèmentaire an X," *Monatsbericht der Gesellschaft zur Förderung der Wissenschaften, des Ackerbaues und der Künste im Unter-Elsass gegründet 1799/Bulletin mensuel de la Société des Sciences, Agriculture et Arts de la Basse-Alsace fondée en 1799* 24 (1890): 8–51, here 37.

58 In the *Mercure du département de la Roër*, an anonymous author even published an ode on his work in rhymes in 1811: Anonymous, "Impromptu fait après lu l'instruction de M. Cadet-De-Vaux sur un nouvel engrais pour la culture du tabac," *Mercure du département de la Roër* 2 (1811): 636.

59 Directeur général de l'Administration des Droits Réunis to Lezay-Marnésia, Paris, 11 February 1812, ADBR, 11/M/196.

60 Louis Metzger, *Coup d'oeil sur la culture, la fabrication et le commerce du tabac en Alsace aux XVIIe et XVIIIe siècles* (Strasbourg: Impr. des "Dernières nouvelles de Strasbourg," 1938), unpaginated.

61 Regarding the director of the manufacture Ange-Marie Francois Gaétan Marocco, Michel Richard and Roger Dufraisse, *Grand notables du premier empire. Bas-Rhin, Sarre, Mont-Tonerre, Rhin-et-Moselle, Roer* (Paris: Éditions du Centre National de la Recherche Scientifique, 1978), 27.

62 Letter, Inspecteur général des Droits Réunis to Préfet du Département de la Roer, Cologne, 12 March 1812, Landesarchiv Nordrhein-Westfalen (Duisburg) (LA NRW), Roerdepartement, 02023, Fol. 249–251.

63 Letter, Inspecteur général des Droits Réunis to Préfet du Département de la Roer à Aix-la-Chapelle, Cologne, 13 June 1812, LA NRW, Roerdepartement, 02023, Fol. 324.

64 For a similar argument, see Rowe, *From Reich to State*, 104–105: "The financial administrations remained predominantly 'French';" for a more balanced view on this point, Clemens, "Verwaltungseliten," 76–77.

65 Pierre-Yves Lacour, "Les commissions pour la recherche des objets d'arts et de sciences en Belgique, Allemagne, Hollande et Italie. 1794–1797. Des voyages naturalistes?," in Nicolas Bourguinat and Sylvain Venayre, eds., *Voyager en Europe de Humboldt à Stendhal. Contraintes nationales et tentations cosmopolites 1790–1840* (Paris: Nouveau Monde Éditions, 2007), 21–39, here 30–32.

66 Letter, Gmelin to Landes Ökonomie Departement, Karlsruhe, 14 March 1811, GLA, 236/6072, Fol. 12–14.

67 Frank Engehausen, "Die Integration der kurpfälzischen Gebiete in den badischen Staat," in Armin Kohnle, Frank Engehausen, and Frieder Hepp, eds., *...so geht hervor ein 'neue Zeit'. Die Kurpfalz im Übergang an Baden 1803* (Ubstadt-Weiher: Verlag Regionalkultur, 2003), 233–246, here 235; Gantet and Struck, *Revolution, Krieg und Verflechtung*, 106.

68 Letter, Ministre des Finances to Lezay-Marnésia, Paris, 23 November 1811, ADBR, 11/M/196, unpaginated; see also Français de Nantes to Lezay-Marnésia, Paris, 9 October 1811, ADBR, 11/M/196, unpaginated.

69 However, this also revealed the limits of Parisian authority: for instance, on one occasion, seeds from Holland and Flanders, which the central administration had promised to the Prefect of Strasbourg, could not be obtained, so Parisian officials sent instead some varieties from the "midi" of France and assortments from the Départments of Lot-et-Garonne and Ille-et-Vilaine. See: Letter, Directeur général de l'Administration des Droits Réunis to Lezay-Marnésia, Paris, 9 March 1812, Archives départementales du Bas-Rhin (ADBR), 11/M/196, unpaginated. Seed deliveries from the central administration in Paris are also found in the period after 1815: Gustave Fornier de Saint-Lary, *Observations de M. le Directeur général des Contributions indirectes sur les faits exposés par M. Fornier de Saint-Lary, dans son rapport à la chambre des députés. Sur le projet de loi relatif au tabac* (Paris, 1819), 13.

70 Français de Nantes to Lezay-Marnésia, Ibid.

71 The central administration allowed Lezay-Marnésia to publish his tobacco handbook, which had initially been distributed only to farmers and scholars in Alsace, in the *Annales de l'agriculture françoise* [sic]: Adrien de Lezay-Marnésia, "Manuel du cultivateur de tabac. À l'usage du département du Bas-Rhin," *Annales de l'agriculture françoise, contenant des observations et des mémoires sur toutes les parties de l'agriculture* 49 (1812), 58–95.

72 Cited in Egon von Westerholt, *Lezay Marnesia. Sohn der Aufklärung und Präfekt Napoleons (1769–1814)* (Meisenheim am Glan: Verlag Anton Hain Meisenheim, 1958), 209, who mentions "German journalists" that addressed Lezay-Marnésia.

73 André Cottez, *Un fermier général sous le Consulat et l'Empire* (Paris: Librairie du Recueil Sirey, 1938), 218.

74 E. Gondolff, *Le tabac sous l'ancienne monarchie. La ferme royale 1629–1791* (Vesoul: Cival, 1914), 215.

75 Kenneth Margerison, "P.-L.- Roederer: Political Thought and Practice in the French Revolution," *Transactions of the American Philosophical Society* 73 (1983): 1–166, here 42–54.

76 Michel de Truchet, *Mémoire sur la nécessité d'étendre la culture du tabac en France, pour éviter l'exportation du numéraire; et sur l'examen analytique des tabacs français, d'après lequel on peut avoir l'assurance de trouver en eux les qualités nécessaires à la bonne fabrication* (Paris, 1816), 70–71.

77 Vigié and Vigié, *L'herbe à Nicot*, 187 and 344–346.

78 van Wickeren, *Wissensräume im Wandel*, 80–81.

79 Alexis-Antoine Cadet de Vaux, *Traité de la culture du tabac, et de la préparation de sa feuille, réduites à leurs vrais principes* (Paris, 1810), i.

80 Peter Sahlins, "Natural Frontiers Revisited: France's Boundaries since the Seventeenth Century," *The American Historical Review* 95 (1990): 1423–1451.

81 Carl Ludwig Lokke, "The French Agricultural Mission to Egypt in 1801," *Agricultural History* 10 (1936): 111–117, here 115.

82 The most important examples are sugar beet and cotton: Joseph Horan, "Napoleonic Cotton Cultivation: A Case Study in Scientific Expertise and Agricultural Innovation in France and Italy, 1806–1814," in Sharon Kingsland and Denise Phillips, eds., *New Perspectives on the History of Life Sciences and Agriculture* (Cham: Springer, 2015), 73–91; Emma C. Spary, *Feeding France: New Sciences of Food, 1760–1815* (Cambridge: Cambridge University Press, 2014), 268–315.

83 Cited in Franz, *Johann Nepomuk Schwerz*, 10–11.

84 The example of Schwerz also shows that the assumption of Gabriele B. Clemens, that administrative experts did not use "national arguments" at all during the Napoleonic era and that nationalism only played a role in the Rhineland after the 1840s, is untenable: Clemens, "Verwaltungseliten," 89–90.

85 Michael Jeismann, *Das Vaterland der Feinde. Studien zum nationalen Feindbegriff und Selbstverständnis in Deutschland und Frankreich 1792–1918* (Stuttgart: Klett-Cotta, 1992), 27–152.

86 Cited in Boerner, *Kölner Tabakhandel*, 102.

87 Bernhard Struck, *Nicht West – nicht Ost. Frankreich und Polen in der Wahrnehmung deutscher Reisender zwischen 1750 und 1850* (Göttingen: Wallstein, 2006), 193–229.

88 Letter, Directeur générale to Préfecture du Département du Bas-Rhin, Paris, 14 March 1815, SK, N5/Nachlass Johann Nepomuk Schwerz, unpaginated.

89 "Das linke Rheinufer; des Stromes Rand bis zum Stolzenfels," in *Denkwürdiger Rheinischer Antiquarius, welcher die wichtigsten und angenehmsten geographischen, historischen und politischen Merckwürdigkeiten des ganzen Rheinstromes [...]* (Koblenz, 1831), 213–244, here 242; in the Bas-Rhin, the long-standing esteem in which the prefect's tobacco reform was held was demonstrated by the fact that a statue of Lezay-Marnésia was inaugurated in August 1857 in Strasbourg, which was attended by a representative of the Grand Duchy of Baden: see von Westerholt, *Lezay Marnesia*, 226.

90 F. E. Foderé, "Extrait d'un Mémoire manuscrit sur la Culture du tabac dans le Département du Bas-Rhin, lu à la Société des sciences, agriculture et arts, par M. J. B. N. L. Husson, employé supérieur de l'Administration des contributions indirectes," *Journal de la Société des Sciences, Agriculture et Arts du Département du Bas-Rhin* 1 (1824): 226–257.

4 Global Communication and Construction of Knowledge in French Naval Medicine

Pierre-François Kéraudren and the Health Department of French Navy, 1813–1845

Daniel Dutra Coelho Braga

Introduction

In early July 1832, a decision made in Paris may have deeply upset French Navy officer Joseph-Fortuné-Theodore Eydoux (1802–1841). Pierre-François Kéraudren (1769–1857), then Inspector General to the Health Department of French Navy, decided that Eydoux, who had just accomplished a circumnavigation expedition as major surgeon,[1] would not be promoted in rank within the French Navy. It was also settled that the surgeon from Toulon would not be allowed to move to Paris in order to organize the display of natural history collections that he himself had painstakingly assembled during his journey around the world. In the inspector's words, Eydoux should have "at least the right to a leave in order to recover from his tiredness," but, even though the surgeon had kept sanitary protocols throughout the travel with considerable "zeal" and had shown "enlightened views" on measures to be taken on board, he should by no means move to the French capital.[2]

At first glance, such a disagreement between Kéraudren's and Eydoux's expectations might seem to have been nothing but an ordinary clash between people who, despite belonging to the same institution, stood at different hierarchical levels. Kéraudren denied career advancement to Eydoux in the same way that he may have done to other officers. However, differences between Eydoux and Kéraudren – himself a French Navy officer[3] – went beyond military ranks. Each officer symbolized a particular position in the evolving process of construction of knowledge that their institution was eager to attain. Not only would they represent disparities between a subordinate officer and a physician at the head of the Navy Health Department, but they would also embody the contrast between travel and cabinet, practice and protocol, experience and synthesis, and, ultimately, between movement and permanence.

Those disparities notwithstanding, Kéraudren and Eydoux remained connected through a basic, though complex, process: communication. As an inspector based in Paris, Kéraudren had to assemble and analyze correspondence exchanged between several officers, who were acting in the same way as

DOI: 10.4324/9780367814540-5

Eydoux: traveling, collecting, healing, and quite often harshly surviving overseas. Communication was thus a basic, if not the most important, component of Kéraudren's ordinary tasks and missions. Because of that, the foregoing contrasts embodied by both were much more entangled than they may seem at first glance, especially because communication would play a decisive role in the construction of knowledge.

Indeed, communication is a key aspect of constructing knowledge.[4] Consequently, a thoughtful approach to the many strategies, layers, and chains of communication that enable construction and validation of scientific formulations may unveil several features within the production of knowledge, such as controversies and disputes among those who took part in it, as well as underlying assumptions that might have been effaced in final syntheses. Moreover, the notion of communication addresses texts, which can be historically analyzed not only in terms of their differences regarding genres, but also in terms of editing and publishing.[5] The impact of editing on conditions of meaning and transmission of texts[6] plays a decisive role in the circulation of knowledge. Finally, in the face of challenges that have been increasingly posed by scholarship on the global aspects of science,[7] the problem of the potential and limits of global communication in the construction of knowledge within the French Navy institutional framework arises. And this is the case especially in regard to French naval medicine, which has been inherently framed through the experience of colonial spaces.[8]

That being said, this chapter explores the role of communication within activities of the French Ministry of Navy and Colonies that were essential to the construction of French naval medicine. My contention is that continuous communication between Kéraudren and other officers, either travelers or physicians based in colonial spaces, helped maintain a multiple-way institutional circuit of production of knowledge, which enabled the construction of increasingly complex medical categories. I argue that the French Navy Health Department based in Paris was able to address the challenges of overseas management, long-distance interlinkages, and local diseases not as center, but rather as an inextricable communication knot. It managed to construct knowledge through the maintenance of concomitant processes of cross-border communication and, by doing so, it kept the French Navy from sustaining reductionist medical categories. Therefore, Kéraudren was neither a mere center of calculation nor a receptacle for traveling officers' accounts, but a communication node that attained a specific standpoint in the construction of knowledge, by assembling, evaluating, and editing scientific formulations produced outside Paris. In such a process, various kinds of text, with their own specific functions and formats, played different roles in the construction of knowledge. This very effort of communication through different texts, I suggest below, enabled a complex and ongoing construction of medical categories regarding the so-called "warm climates," due to which variables such as what Navy officers called "maritime cities," the spatial arrangement of ships, and even watercraft movements themselves ended up being as crucial to French naval medicine as divisions of the globe in isothermal zones.

Within and without a Center: the French Navy's
Globality and Institutional Framework

The French Navy was global in scope, and so was French naval medicine. The nature of such globality, however, is not self-evident. Both the French Navy and French naval medicine can be seen as global mainly because of French colonial patterns and the steady maritime activities which constituted them, and yet this globality should not be automatically conceived as evidence of effective power. Consequently, attention must be paid to the words used to describe it. While global French territoriality at the time was much more an envisioned aim than an actual premise, a territorial globality, even though harshly kept through the maintenance of territories in different continents and oceans, allowed an ongoing set of interlinkages that ensured transformations, not only of French overseas territories, but of France herself.

Some of these transformations took place within scientific fields. The French colonial realm at the time when Kéraudren led the Health Department has been described by historians in diverse ways, especially because of its connections to slavery, specific environments, indigenous populations, and transnational politics.[9] Nevertheless, however divergent interpretations regarding French colonial guidelines at that time may be,[10] the maintenance of overseas territories ensured transcontinental interlinkages, which played a key role in the construction of knowledge. Such maintenance was ensured by a French Navy that, in spite of all the political language that insisted on reappraising the value of an alleged "*Ancien Régime,*" had presented modern characteristics since the previous century.[11] The post-Napoleonic French Navy was imperial and colonial in its guidelines and yearnings, and so were their scientific activities. A remarkable example of those yearnings were scientific expeditions around the world carried out by the French Navy.[12] Moreover, efforts within the French Navy in the realm of scientific publications were quite obvious, a case in point being the regular edition of an institutional journal, the *Annales maritimes et coloniales*, which presented a volume solely devoted to "science and arts" since its first release.

If the French Navy fed imperial and colonial yearnings through science, its scientific activities, on the other hand, mostly contributed to changes in France herself, which prompts a more complex picture of relations between metropolitan and colonial spaces in terms of science and construction of knowledge.[13] A noteworthy aspect of those changes were the differences in terms of scientific practices and formulations that could be recognized between French institutions based in Paris and those based in port regions, which benefited from connections inherent to port life.[14] This picture leaves the possibility of identifying a center for all this framework quite ambiguous, even though it is true that the French Navy, due to specificities of the French political administration, had an epicenter where communication would be centralized: the French capital itself.[15]

Paris' centrality in terms of political, administrative, and even communicative networks is evident. However, Ministries based there, as well as some arenas of representative politics – such as legislative chambers –, were under constant tension, which would eventually prevent this centrality from being as "inertially diffusionist" as some descriptions of a center may often conceive it. Paris was a rather unstable place for the French Navy. Scholars have even pointed out the instability which characterized the French Ministry of Navy and Colonies throughout the monarchical regimes due to recurrent change of Ministers.[16] At first glance, such frequent change of Ministers could denote an equally frequent transformation of the whole institutional framework. However, there were more subtle nuances within this picture. It is true that several positions were mostly attached to the Minister himself, as it was the case of the "*sous-secrétaires d'État*," who were chosen by Ministers once they were within the French state.[17] However, the French Navy's institutional framework was structured through several positions, such as the Direction of Ports, the Maritime Prefectures, and the aforementioned Health Department, among others. Although positions were hierarchically subordinated to the main Minister, one could wonder to which extent officers who remained in their positions for longer did not manage to achieve a certain level of power and control that made them more efficient and authoritative – at least locally – than much of what was obtained by a Minister.[18]

This seems to have been the case of the Health Department of the French Navy in Paris, and it is within this realm that one may grasp the meaning of Kéraudren's permanence in the position of inspector general, from 1813 to 1845. The department itself benefited from a degree of stability which other channels did not benefit from. Kéraudren's permanence was, therefore, not a sign of inertial action. He was fairly active, and an essential part of his actions were linked to writing.

Constructing French Naval Medicine through Texts

An impressive aspect of Kéraudren's scope of actions within the French Navy Health Department was his writings. He produced a wide range of texts, which varied from instructions regarding voyage protocol to memoirs and articles, which ended up being published in the journal *Annales maritimes et coloniales*. Despite their differences, all the authorial acts which resulted in the production of these texts benefited from the appropriation of texts produced through a global communication framework.

A case in point is a specific set of travel instructions, which focused on the hygiene protocol to be observed onboard. Within the realm of travel instructions as a textual genre,[19] sanitary instructions stood out due to their own scope of practices, topics, and protocols.[20] A noteworthy piece of evidence of Kéraudren's role in voyage protocol is the manuscript of sanitary instructions that he wrote in regard to Louis de Freycinet's expedition.

This manuscript, written in Paris, consists of almost 30 pages and was dated 1 January 1817.[21] It is composed of two main sections, the first one dedicated to "dispositions regarding ship cleanliness and the crew's health" ("*dispositions relatives à la salubrité du vaisseau, et à la santé de l'équipage*"), while the second one consisted of "remarks on salting" and mostly focused on the food cleanliness protocol to be followed throughout voyages. In spite of having been conceived in order to fulfill a prescriptive duty, Kéraudren's instructions emphasized conceptual choices and intertextual references. They mostly addressed the problem of humidity within the ship, focusing on the description of devices such as ventilation fans and ovens, which could be useful to maintain proper air conditions inside the watercraft. What was mostly emphasized was ship management in accordance with the sanitary protocol that underlined the quality of air, which should be constantly "renewed" and managed not only through mechanical actions, but also through purification based on chemical management, particularly through the use of the then called "muriatic acid gas." The manuscript also emphasized avoidance of problems such as rats, which were likely to proliferate onboard. Most remarks on ship management were reinforced through references to previous travel writings, such as reports by Captain James Cook.[22]

Kéraudren's sanitary instructions assigned to Louis Duperrey in 1822 differed slightly from those written in 1817 relative to Louis de Freycinet's expedition. The text written in 1822 was shorter and complementary to the instructions previously assigned to Freycinet. The inspector had attached a copy of his own memoir on the causes of diseases on board – which we will analyze below –, as well as a sanitary guide to voyages in Africa. The instructions insisted upon analyzing the causes of diseases on board.[23] This is evidence of how naval hygiene remained an issue to be continuously evaluated and eventually transformed, and not merely applied or strictly followed as a bureaucratic procedure. Watercraft space would be lived as a laboratory, and Kéraudren himself used this word in his manuscript in order to specify procedures regarding remedies control: "a small laboratory shall be established in a part of the vessel, where the most common medicines shall be stored, and where the pharmacist will prepare the remedies prescribed daily."[24]

This aspect of handwritten instructions encourages a more complex reflection upon the medical categories that structured them. In these writings, Kéraudren used expressions such as "torrid zone" and "warm climates." However, these words were not used in a rigid conceptual framework. Their occurrence is not evidence of a so-called "pre-Mansonian tropical medicine,"[25] and the search for evidence supporting that assumption may lead to a false dilemma. As a matter of fact, those words reveal an effort of continuous evaluation of experiences in different places. They were an invitation to localist observations, since vessels themselves were a place where harsh divisions between temperate and tropical zones might fade away. It is plausible to presume that these expressions may have had a quick effect in terms of protocol communication among French Navy officers – particularly through manuscript communication –, but they do not reveal the underpinnings of ultimate conceptual choices.

Indeed, the fact that the ship itself became an essential place for the eventual validation and construction of knowledge evidences the way that French naval medicine categories were continuously open to revaluation. It was not merely a place where Navy officers would obey sanitary protocol, but a place where they would persistently reevaluate their own practices and reflect upon causes of diseases. Kéraudren left French navy travelers free to appropriate the ship as a space of construction of knowledge by suggesting a protocol of naval hygiene that demanded an ongoing examination of everyday practices.

Louis Duperrey, for instance, left traces of how naval hygiene was followed throughout his expedition in a rather active way. In late October 1822, during a stay in the province of Santa Catarina, in Brazil, Duperrey described some of the practices carried out on board in order to preserve his crew's health:

> Having crossed the Cape Verde islands, the sea beautified, though the sky became too clouded. Since intense heat began to hassle us, I resolved that sailors should take their baths alternatively in the bathtubs taken at Toulon for this aim; that Tenerife wine would be drunk every other day instead of liquor and that coffee would offer twice a week during our passage on the tropics. I have equally taken all proper measures in order to prevent the steerage from being affected by the extreme humidity that reigns in these climates, and I recognize with satisfaction that the kitchen and the oven that I had placed on that part of the ship contribute to fulfill this condition of hygiene so much recommended by naval hygiene.[26]

Even though Duperrey did not explicitly mention the name of Kéraudren on his letter, it is evident that he had the inspector general's instructions as the main reference to coordinate the ship space and evaluate tropical climates. Moreover, this letter shows how this evaluation should be communicated to Paris as soon as possible. Since Duperrey had left Toulon in August, the fact that he was promptly providing some feedback on naval hygiene during one of his first stays shows to what extent ongoing communication was a key aspect of French naval medicine. Besides, such an aspect can be traced not only through letters sent to Paris, but also through the travel report itself, despite its peculiar conditions of publication.[27] A brief passage of it shows how notes written at the port of the island of Santa Catarina were important enough to be provided in the printed version of this travel account:

> We had been experiencing for some time the nuisances of the first tropical heats. As we were not used to them, the sensation was even more significant. Temperature remained the same both night and day, between 25 and 28 degrees Celsius. In order to attenuate indispositions that could have resulted from this fact, we had bathtubs installed so that the crew could bathe every day, under the supervision of Mr. Garnot and Mr. Lesson, who did it with particular attention. We chose to offer Teneriffe wine at breakfast, instead of liquor, since it was less exciting. We also took all proper measures in order to keep the watercraft from extreme humidity that characterizes equatorial seas; and, since fire is one of the most powerful means to purify the altered air of vessels, we were pleased to realize that the kitchen and the oven, that we had settled in the steerage, perfectly met that condition of cleanliness particularly recommended in Mr. Kéraudren's naval hygiene.[28]

There are telling differences between both passages. The erasure of the reference to Toulon in the final version of the account, for instance, may have reinforced a representation of French port life as less decisive than it had actually been.

The fact that bathtubs had been prepared in Toulon, nonetheless, shows how French port sites were responsible for the actual implementation of naval hygiene in its most practical demands, especially regarding spatial management of vessels. Finally, those differences show how the chain of communication within the French navy codified not only hygiene protocol, but even Kéraudren's authority itself, which was mentioned on the pages of the printed account.

Kéraudren as a Published Author

Kéraudren also articulated his conceptual guidelines in published memoirs, notably *Memoir on the causes of diseases of seamen and measures to be taken in order to keep their health at ports and on the sea,* whose first edition came out in 1817, as well as *On yellow fever, observed in the Antilles and on Royal vessels, taken into account mostly in regards to its transmission*, published for the first time in 1823.[29] Both memoirs were published by the Royal Printing Works. The fact that his thesis on scurvy had been brought out in 1804 through a private publishing house[30] illustrates to which extent this new writing phase was also the result of an institutional framework in which communication, the construction of knowledge, and colonial guidelines were interwoven.

The first edition of the *Memoir on the causes of diseases of seamen* consisted of about 100 pages divided in three sections. The first one is dedicated to healthiness maintenance on ships, while the second one refers to broader discussions on "the physical and moral state of men on the sea." Finally, the third section adds spatial variables to the debate, by addressing "the health of sailors throughout coastal navigation and ship halting." In the same way as the memoir *On yellow fever*, Kéraudren's *Memoir on the causes of diseases of seamen* was conceived through wide intertextuality. The most recurrent source was a treatise written by Dutch physician Louis Rouppe in the eighteenth century, which had also been a main intertext for the inspector general's thesis on scurvy.[31] Another reference was an article published by doctor Wuettig, in which an air purifier was thoroughly described, as well as a treatise on means of disinfecting air, written by Guyton-Morveau and published in the *Annales maritimes et coloniales*.

Travel writing, however, remained the main intertextual foundation. Several examples derive from the *Voyages de Stavorinus*, a chief officer who had served the Batavian Republic, and the voyage of Captain Cook was the most outstanding reference. Not only was Cook's journal a recurring source for Kéraudren, but texts written by Cook's entourage were also appropriated by the inspector. Hence, writings by doctor Pringle, then Royal Society president, were framed in Kéraudren's treatise as textual cornerstones. Writings by other travelers, such as Alexander von Humboldt, Aimé Bonpland, Jean-François de Lapérouse, and even François Péron, who followed Nicolas Baudin's expedition, were also used. Finally, a brief reference is made to French physician Alexandre Moreau de Jonnès, who had analyzed seamen's health in the Antilles at that time.[32]

One might wonder to which extent Kéraudren's references for this memoir remained European rather than global. Nevertheless, all texts recovered by Kéraudren had been derived from transcontinental experiences and previous global communication exchange. In addition, in the hidden intertextuality of his text one may notice broader hints and clues of global communication and transnational circulation of knowledge. Kéraudren's early handwritten sanitary instructions, for instance, were one of these hidden intertexts. The published memoir was more than a printed edition of the notes assigned to Freycinet, but connections between both texts are crystal clear.

However, what better indicates the impact of global communication in construction of knowledge is the textual and chronological gap between the first edition of the published memoir and a second one, published in 1824.[33] The second edition benefited from actions carried out by the French Navy in a global scale, such as the maintenance of a dynamic port life in France and, most of all, two scientific voyages around the world.

Between 1817 and 1823, Freycinet successfully achieved his expedition around the world, while Duperrey promisingly carried on his own circumnavigation. These expeditions remained linked to Kéraudren, since those officers and the surgeons who accompanied them had to write thorough reports commenting diseases analyzed throughout the expeditions and the sanitary state of their groups. Even though they had been mostly coastal and insular in their itineraries – a characteristic that led geographer Numa Broc to describe their officers as "maritime voyagers" and not actual explorers[34] –, those expeditions envisioned a sense of globality and provided material for a transnational circuit of medical reflection. In addition, several French merchant ships also achieved circumnavigations. Those voyages were not conceived following usual French Navy standards, but they ended up being quite meaningful. An outstanding expedition was the one commanded by Camille de Roquefeuil, who left Bordeaux in October 1816 and returned to France in November 1819.[35]

The second edition of *Memoir on the causes of diseases of seaman* also referred to French port life. This can be recognized through a rather small, though meaningful, piece of evidence: a footnote that mentions the port of Toulon. This footnote addresses activities carried out by Édouard Burgues de Missiessy (1756–1837), a port commander whose trajectory indicates stability within French Navy positions. A brief mention of a letter received by Kéraudren and written by Missiessy himself shows not only the impact of the inspector general's sanitary guidelines on voyage practices, but also the availability of information produced during voyages, which had their path "from globe to Paris" ensured by port life actors such as Missiessy himself, as we can see in the way Kéraudren quoted this piece of correspondence:

Since the first edition of this Memoir, this piece of advice was executed on several Royal vessels. Here is a message Mr. Vice-admiral, the count Burgues-Missiessy, Navy commander in Toulon, has honored me with on this topic: "your views on the installation of kitchens and the oven have never ceased to be the same as mine. Without even mentioning the extent to which these installations would be favorable to stability, their enormous advantage in regard to sailors' health would be that of being the best means of keeping the inferior parts of vessels dry and clean, where humidity is so constant and pernicious, as well as [the advantage of] offering sailors exposed to cold or rain the possibility of warming themselves. It was with this intention that vessel kitchens were placed in the steerage of the vessels *le Trident* and *le Scipion*, recently armed, as well as in the frigates *la Guerrière, la Médée* and *la Thétis*. The corvette *la Coquille*, which is at the moment on a circumnavigation voyage, had them installed in this way as well, and, as you already know, lieutenant Duperrey, who boast himself on the health of his crew, attributes this achievement mostly to the position of kitchens, even though his steerage is rather low."[36]

An ongoing evaluation of sanitary protocol through spatial management of ships at ports was also enabled through the foregoing expedition commanded by Roquefeuil. Reports derived from this expedition were available to Kéraudren, since they had been published on the *Annales maritimes coloniales*, besides having had a final report published, which explains how the inspector based in Paris was able to use Roquefeuil's example in the second edition of his memoir, drawing attention to the necessity of using sulfurous acid gas to clean ships once they have disembarked at ports. According to the inspector, this measure was effective against the spread of rats on a vessel, even though rodents could often survive it. Nevertheless, Roquefeuil's successful use of sulfurous acid gas against rats in the port of the Island Kodiak, in Alaska, was recovered by Kéraudren in order to reinforce the measure.[37]

Finally, another noteworthy aspect of global communication incorporated into the second edition of the *Memoir on diseases of seamen* was a letter written by a lieutenant identified as Fleurieu. Entirely reproduced as an appendix,[38] this piece of correspondence describes movements carried out on the vessel *La Pomene* from the Antilles to several ports in South America, such as those in Rio de Janeiro, Santa Catarina, and Lima. Kéraudren made use of it in order to reinforce his own arguments regarding the use of specific ventilation devices within vessels. Though the letter was written in Paris, on 24 February 1824, it is undoubtedly the result of an enterprise characterized by ongoing communication within the French Navy's institutional framework, as evidenced by the many editions of the *Annales maritimes et coloniales* which have described the movements of Fleurieu overseas.

All those examples notwithstanding, it is the memoir *On yellow fever* that reveals the widest connections enabled by global communication. The text consists of two main parts, preceded by a dedication to then Ministry of French Navy and Colonies, Marquis de Clermont Tonnerre, and an epigraph in Latin. The first part of the "*mémoire*" depicts the perspective according to which yellow fever would never be contagious, while the second part of it analyzes "new facts regarding yellow fever transmission." Due to his position as an authorial assembler benefiting from global communication channels, Kéraudren was able to insist on a balanced position toward the controversy, claiming a less radical approach to the disease's etiology. In his own words,

It is on the tendency that men usually have to claim exclusive opinions that we must seek the cause of the current disagreement among physicians; as if, once circumstances have changed, results could remain the same. Some say: yellow fever was not contagious in this and that case, therefore it never is so. Others answer: on this and that occurrence, yellow fever was evidently contagious; and so it always is. Both propositions are equally based on facts, but for one as for the other the conclusion is way too general and way too absolute.[39]

Such a standpoint was attained through wide intertextuality, knotted through several text genres. One of the main textual interlocutors of Kéraudren was the Royal physician and French navy surgeon Pierre Lefort (1767–1843), cited through different sources. Lefort's own "*mémoire*" and one of his letters that ended up being published in the *Journal universel des sciences médicales* were important sources.[40] The paramount reference of Lefort notwithstanding, reports written by major surgeons who had been traveling on royal vessels were the most frequent kind of text appropriated by Kéraudren. The inspector general even highlighted their position as "traveling physicians" when analyzing their perspectives. Surgeons like Aubert, who had served as the main surgeon on the frigate *la Nereide*, Bonet, who served as a surgeon on the corvette *la Gloriole*, and Jolivet, a surgeon who embarked on the royal frigate *l'Africaine*, were as important as Lefort in terms of textual references. Kéraudren also made use of "*compte-rendus*" sent to port health councils in France. This was, for instance, the case of the text written by Péan, a surgeon who had written to the Health Council of the port of Brest. Péan's text was appropriated as a counterpoint to some of Lefort's assumptions. Not only does this evidence the connection between Kéraudren and French port life, but it also shows to which extent handwritten reports and *compte-rendus* were as important as printed memoirs and articles on journals. There was no previous hierarchy among textual genres themselves when it came to analyzing yellow fever. Notes written on board could be as insightful as a "*mémoire*" or a letter that had already been edited in a journal, and therefore received thorough attention from the inspector based in Paris.[41] The construction of French naval medicine through global communication was based on a myriad of text genres, in which publication hierarchies could be temporarily suspended to enable thoughtful appropriation.

This aspect was even highlighted by Kéraudren himself, who drew attention to the occurrence of what might have been one of the most remarkable examples of communication between the French Navy Health Department in Paris and physicians settled in the Antilles. In the first part of *On the yellow fever*, which argued that the disease was not contagious, the inspector general pointed out that "physicians in Martinique and Guadeloupe, consulted in 1819 by Your Highness, the Navy Minister, on the question of knowing whether, in their opinion, the yellow fever was contagious or not, had largely opted for the negative [answer]."[42] In this passage, Kéraudren was making a reference to a thoughtful request made by the French Navy regarding recurrent cases of yellow fever in 1816. This request was successful thanks to the action of several men, including François Xavier Donzelot (1764–1843), then governor of Martinique, and Edme Michel Mauduit (1773–1848), then head of the Colonial Bureau, in Paris.[43]

In February 1820, Donzelot gathered twenty-one reports written by physicians based in Martinique and sent them to the Health Department of the French Navy in Paris. In a letter sent to the head of the Colonial Bureau, Donzelot highlighted that he had consulted not only physicians based at colonial hospitals, but also surgeons who had had longer practice time in different colonial places.[44] Three months later, those letters were forwarded to Kéraudren by Mauduit, who urged the inspector general to "share with the Colonial Bureau, as soon as possible, the observations that shall derive from the exam of these memoirs."[45] Mauduit was ensuring the efficiency of a communicative chain that had also involved the then Navy Minister Portal, who had been the first addressee and main interlocutor of Donzelot, who sent the physicians' reports to Paris along with the following note:

> After having received the letter n. 118 that you did me the honor of writing to me on 28 April 1819, relative to the sanitary state of the Colony, I communicated the note which it contained, written by Mr. Kéraudren, not only to the chief health officers of the hospitals, but I took the opportunity regarding the material which it deals with, to probe the opinion of the doctors and surgeons having a more or less long practice and who are known to practice with some distinction in the various districts of the colony.[46]

Previous research has already shown to which extent the French Antilles were a place of continuous construction of knowledge.[47] What this exchange of correspondence between men like Donzelot, Mauduit, and Kéraudren shows is that not only did the Antilles remain an important place of construction of knowledge throughout the first half of the nineteenth century, but they were also deeply interwoven to metropolitan formulations due to continuous and efficient communication.

Another noteworthy example of reports assigned to Kéraudren was the one written by the aforementioned officer Eydoux, as a result of his participation in the expedition commanded by Cyrille Laplace.[48] The report that eventually arrived at Kéraudren's hands had been written in Toulon. It was addressed to the navy Health Council of Toulon and subsequently forwarded to Laplace, who transmitted it himself to the Navy Minister.[49] Even though Eydoux's texts[50] were not cited by Kéraudren, this example of a medical report transmission chain provides a picture of how the inspector general continuously received new input provided either by surgeons or by commanders who had been traveling overseas. Therefore, it is plausible to infer that even his early instructions had been based on similar material.

Finally, Kéraudren's authorship and its connections to global communication can also be visualized through his frequent contributions to the *Annales maritimes et coloniales*. Even the aforementioned memoirs by the inspector general were conceived through direct mention to his activity in that journal as an author. The first edition of the memoir on diseases of seamen, for instance, cited his own article on maritime atmosphere and seawater, which had been published in 1816,[51] while its second edition was entirely published in the 1824 edition of the *Annales*.[52] On the pages of this journal, the inspector general himself pointed out to what extent his authorial productions depended upon letters and reports that he received through the French navy's own global communication channels. When addressing diet protocol on board, for instance, in an article,[53] Kéraudren referenced experiences carried out by Albin Roussin,[54] a French Navy officer who had been acting on naval stations and on his own expeditions in South America since the early years of the newly restored monarchy.[55]

The *Annales Maritimes et Coloniales* have also played a role in the maintenance of controversies that were intrinsically global, such as the one regarding yellow fever. It was in the journal's pages that Kéraudren was able to once again counter the positions held by Lefort. As it was mentioned before, even though Kéraudren had used Lefort's account in his memoir on yellow fever that was published in 1823, this did not mean that the inspector general based in Paris agreed with most of the physician's views on the nature of that disease. Hence, Kéraudren used the *Annales maritimes et coloniales* as a platform where he was able to assertively counter Lefort's view on the presumably noncontagious nature of yellow fever. He did it by publishing the note *On yellow fever considered in regard to Europe and France, current state of the question*, in 1824. In this note, Kéraudren affirmed that "it would be too dangerous to adopt, with all its consequences, the opinion of those who believe that this disease cannot be communicated through contagion." Once again, Kéraudren managed to reply to an interlocutor whose field was rather local – since Lefort restrained his assumptions to Martinique only – by appropriating the views of another author whose observations had been structured through displacement and movement. The main source for the new reply by Kéraudren were the notes written by the physician Rochoux, who had worked on yellow fever flows not only in the Antilles, but also in Southern Spain.[56] Since Rochoux had not been mentioned by the inspector in the memoir *On the yellow fever,* published just one year before, it appears that new formulations on the disease in cities like Barcelona and Cadix, in a thoughtful comparison with the Antilles, had just been communicated to the Health Department, which led Kéraudren to promptly submit an emphatic reply to the most important journal among French Navy circles.

Conclusion

All the foregoing examples indicate that Kéraudren was not merely a cabinet author. He was an authorial assembler, whose position enabled a specific standpoint through which he managed to compare, balance, and validate formulations that remained, nonetheless, imbued with – and even subordinated to – localist references, all of which had movement and place as epistemological cornerstones. Therefore, these examples help us in terms of locating construction of knowledge within the French naval medicine field. It was not only in France that French naval medicine developed itself. The degree to which an ongoing debate on medical categories was kept is evidence that a harsh division between a metropolitan science and a colonial one was not the case regarding the French Navy Health Department.[57] As an author based in Paris, Kéraudren was extremely dependent upon texts written either on ships or on the sea, in colonial spaces or even at French ports. A better understanding of these localities may prevent us from reinforcing a teleological or reductionist conception of what could have been a French centrality in this process. Indeed, the French specificity of centralization of administrative power within French state building should not be considered as evidence of spatial centrality in the construction of knowledge. Centralization within State-building is not the same as centrality in the construction of knowledge. Power and science were intertwined in French overseas territoriality, but their expression in space followed different vectors and guidelines, and Kéraudren's activities are one among many examples of these vectors.

Moreover, Kéraudren's texts indicate deeper aspects of the construction of medical categories within French naval medicine. French navy commanders and surgeons had to be attentive to maintenance of practices, watching carefully and taking notes of actions and phenomena that could ensure the respect of guidelines previously prescribed through written sanitary instructions – or challenge them. Within an institutional framework that ensured a reasonable degree of communication between ships and Paris, written information on everyday practices on the ship enabled those practices to become a key part in the construction of French naval medicine, instead of just a site where rules should be obeyed.

In sum, these examples show how communication plays a role within the circulation of knowledge. As historian Michael Worboys put it, "the form and timing of transfers involved choices,"[58] and it is quite fruitful to think about the extent to which those choices took place in the form of communicative acts. Kéraudren was not a center of calculation. He was a knot of communication. His administrative centrality in the French structure enabled a mode of reassembling and confronting data that derived from a dynamic transnational circuit of experience and exchange. This circuit was as important as Kéraudren's authorial and editorial terrain in Paris, and, without such a circuit, the whole French naval medicine framework would collapse. In order to articulate and construct medical categories, Paris depended on Point-à-Pitre, Saint-Pierre, Brest, and Toulon. Kéraudren depended on Eydoux.

Notes

1 Eydoux joined an expedition commanded by Cyrille Laplace, who circumnavigated the world from late December 1829 to April 1832. Eydoux's trajectory has been described in biographical dictionaries and works on French Navy surgeons. See Bernard Brisou and Michel Sardet, eds., *Dictionnaire des médecins, chirurgiens et pharmaciens de la Marine* (Vincennes: Service historique de la Défense, 2010), 306. For the expedition commanded by Laplace, see Étienne Taillemite, *Marins français à la découverte du monde: De Jacques Cartier à Dumont d'Urville* (Paris: Fayard, 1999), 564–581. A briefer description is also shown on the rather laudatory pages written by French counter admiral Hubert Granier, *Marins de France, conquérants d'Empires, XIXe-XXe siècles* (Rennes: Éditions Ouest-France, 1991), 53–57.
2 Pierre-François Kéraudren, letter dated 4 July 1832, Service historique de la Défense (hereafter SHD), sub-series BB4, codex 1004. All translations to English from French documents were made by the author of this chapter.
3 Kéraudren's trajectory has been described on biographical dictionaries as well. See Étienne Taillemite, *Dictionnaire des marins français: nouvelle édition revue et augmentée* (Paris: Tallandier, 2002), 273; Bernard Brisou and Michel Sardet, eds., *Dictionnaire*

(2010), 440–441. An accurate mention is made by Michael Osborne, *The Emergence of Tropical Medicine in France* (Chicago, IL: University of Chicago Press, 2014), 118.

4   See James Secord, "Knowledge in Transit," *Isis* 95, no. 4 (2004): 654–672.

5   On forms of texts and the processes of their transmission, see Donald F. McKenzie, *Bibliography and the Sociology of Texts* (Cambridge: Cambridge University Press, 1999). For a conceptual reflection upon genres, see Jean-Marie Schaeffer, "De l'identité textuelle à l'identité générique," in idem, *Qu'est-ce qu'un genre littéraire?* (Paris: Éditions du Seuil, 1989), 64–130.

6   See Roger Chartier, "A mediação editorial," in idem, *Os desafios da escrita*, translated by Fulvia M. L. Moretto (São Paulo: Editora UNESP, 2002), 61–76.

7   See, for instance, Sadiah Qureshi, "Science et mondialisation au XIXe siècle," in Pierre Singaravélou and Sylvain Venayre, eds., *Histoire du monde au XIXe siècle* (Paris: Fayard, 2019 [2017]), 254–273; Fa-ti Fan, "The Global Turn in the History of Science," *East Asian Science, Technology and Society* 6, no. 2 (2012): 249–258; Sujit Sivasundaram, "Sciences and the Global: On Methods, Questions, and Theory," *Isis* 101 (2010): 146–158.

8   Osborne, *The Emergence of Tropical Medicine in France*.

9   French colonial realm's surface in 1815 was described as "derisory" by René Remond, *Introduction à l'histoire de notre temps, vol. 2. Le XIXe siècle, 1815–1914.* (Paris: Éditions du Seuil, 1974) 218. For the links between French Navy, French colonies, and slavery, see Robin Blackburn, "French Restoration Slavery," in idem, *The Overthrow of Colonial Slavery, 1776–1848* (1988, reprint, London, New York, NY: Verso, 2011), 473–515.

10 For an analysis which highlights the difference between guidelines from 1815 to 1821, see Marc Michel, "La Colonisation," in Jean-François Sirinelli, ed., *Histoire des Droites en France*, Vol. 3: *Sensibilités* (Paris: Gallimard, 1992), 125–163, in particular, p. 128. A different picture is offered by Paul Claval, "Colonial Experience and the Development of Tropical Geography in France," *Singapore Journal of Tropical Geography* 26, no. 3 (2005), 289–303.

11 An overview on this debate was provided by Michel Vergé-Franceschi, "Marine et Révolution," *Revue historique des Armées*, no. 175, juin (1989): 77–88.

12 For circumnavigations, see Hélène Blais, *Voyages au grand océan: géographies du Pacifique et colonisation, 1815–1848* (Paris: CTHS, 2005); Agnès Beriot, *Grands voiliers autour du monde: les voyages scientifiques, 1760–1850* (Paris: Port Royal, 1962); and Taillemite, *Marins français à la découverte du monde.* Several studies regarding those expeditions remain in the format of unpublished thesis. See, for instance, Catherine Allaire, *L'Expédition Freycinet à bord des corvettes l'Uranie et la Physicienne, 1817–1820: traditions et nouveautés de la circumnavigation française au temps de la marine à voiles* (PhD thesis: École nationale des Chartes, 1989); Géraldine Barron, *Les voyages autour du monde de Cyrille Pierre Théodore Laplace: les campagnes de la Favorite et de l'Artémise* (PhD thesis: École nationale des Chartes, 1998). It is also worth highlighting previous works by Hélène Blais, *Les voyages d'Abel Dupetit-Thouars ou les débuts de l'expansion française dans le Pacifique (1835–1848)* (PhD thesis: École nationale des Chartes, 1996); and idem, *Les voyages français dans le Pacifique: pratique de l'espace, savoirs géographiques et expansion coloniale (1815–1848)* (PhD thesis: École des hautes études en sciences sociales, 2000). I have analyzed French navy expeditions in South American port cities and their subsequent collections and publications. See Daniel Dutra Coelho Braga, *Colonialidade nos trópicos: a América meridional e as viagens de volta ao mundo da Marinha francesa (1815–1852)* (PhD thesis: Federal University of Rio de Janeiro, 2019).

13 See Joseph M. Hodge, "Science and Empire: An Overview of the Historical Scholarship," in Brett Bennet and Joseph M. Hodge, eds., *Science and Empire: Knowledge and Networks of Science Across the British Empire, 1800–1970* (Basingstoke/Hampshire/New York, NY: Palgrave Macmillan, 2011), 3–29.

14   See Osborne, *The Emergence of Tropical Medicine in France*. A pessimistic depiction of French port zones after the Revolution was presented by René Sédillot, *Le coût de la Révolution française* (Paris: Perrin, 1987), 216–219.

15   The centrality of Paris has been a subject of different inquiries. René Sédillot, for instance, argued that the "natural vocation" of Paris was commercial and administrative, and that economic guidelines regarding the city's industrial prosperity from the Revolution onward were due to an "artificial centralization," which would be later backed by "another form of centralization, the one derived from railways." See Sédillot, *Le coût de la Révolution française*, 191–192. Remarks on disparities between the Paris Chamber of Commerce and those of port regions, which would be under the influence of "heirs of the Atlantic economy," were highlighted by Francis Démier, *La France de la Restauration Bourbon (1814–1830): l'impossible retour du passé* (Paris: Gallimard, 2012), 426. On the wider theme of disputes regarding decentralization throughout the nineteenth century in France, see Pierre Rosanvallon, *L'État en France de 1789 à nos jours*, 2nd ed. (Paris: Éditions du Seuil, 1992), 79–84.

16   For details on such instability, see Étienne Taillemite, *L'Histoire ignorée de la marine française* (Paris: Perrin, 1988), 309–310, 318.

17   See François Berge, "Le sous-secrétariat et les sous-secrétaires d'État aux colonies: histoire de l'émancipation de l'administration coloniale," *Revue française d'histoire d'outre-mer* 47, no. 168–169 (1960): 301–376.

18   The baron Jean Marguerite Tupinier, for instance, had been the Director of Ports for more than six years when the revolution of 1830 broke out. See Taillemite, *L'Histoire ignorée*, 317.

19   For the complexity of travel instructions as a textual genre see, among others, Joan-Pau Rubiés, "Instructions for travelers: teaching the eye to see," *History and Anthropology*, vol. 9 (1996): 139–190; Lorelai Kury, "Les instructions de voyage: orienter le regard, former les gestes," in idem, *Histoire naturelle et voyages scientifiques (1780–1830)* (Paris: L'Harmattan, 2001), 91–146. Remarks upon instructions were also made by Kapil Raj, *Relocating Modern Science*, 27–28, 103–104.

20   This aspect may explain why sanitary instructions were not emphasized in the study by Hélène Blais. See the second chapter of the first volume of the thesis written by Blais, "La préparation des voyages: itinéraires et instructions," in idem, *Les voyages français dans le Pacifique*, 105–186.

21   Pierre-François Kéraudren, "Instructions sanitaires," SHD, Sub-series BB4, Codex 999, ff. 21–52.

22   Kéraudren, "Instructions...," *passim*.

23   Pierre-François Kéraudren, "Instruction sanitaire pour les officiers de santé de l'expédition commandée par M. le lieutenant de vaisseau Duperrey," handwritten instructions dated 8 June 1822, SHD, Sub-series BB4, Codex 1000.

24   Ibid.

25   The possibility of identifying foundations of pre-Mansonian tropical medicine was explored by David Arnold, ed., *Warm Climates and Western Medicine: The Emergence of Tropical Medicine, 1500–1900* (Amsterdam/Atlanta, GA: Rodopi, 1996), Introduction.

26   Louis Isidore Duperrey, letter dated October 25, 1822, SHD, Sub-series BB4, Codex 1000.

27   For an analysis which exposes the obstacles and delays regarding Duperrey's publication, see Daniel Dutra Coelho Braga, "O processo de publicação do relato da viagem científica do oficial da Marinha francesa Louis Duperrey," in Tânia Bessone, Gladys Ribeiro, Monique Gonçalves, and Beatriz Momesso, eds., *Imprensa, livros e política no Oitocentos* (São Paulo: Alameda, 2018), 303–334.

28   Louis Isidore Duperrey, *Voyage autour du monde fait par ordre du Roi* (Paris: Imprimerie Royale, n.d.), 23–24.

29   Pierre-François Kéraudren, *Mémoire sur les causes des maladies des marins, et sur les soins à prendre pour conserver leur santé dans les ports et à la mer* (Paris: Imprimerie

Royale, 1817); and idem, *De la fièvre jaune, observée aux Antilles et sur les vaisseaux du Roi, considérée principalement sous le rapport de sa transmission* (Paris: Imprimerie Royale, 1823).

30 Pierre-François Kéraudren, *Réflexions sommaires sur le scorbut, thèse présentée conformément à l'article XX de la loi, et soutenue à l'École de Médecine de Paris, le 9 Pluviôse an 12* (Paris: Lebour, 1804).

31 Kéraudren, *Réflexions sommaires sur le scorbut*, 3–4, 7, 9, 13, 15, 17–18, 30; idem, *Mémoire sur les causes des maladies des marins*, 33, 38–40.

32 Kéraudren, *Mémoire sur les causes des maladies des marins* (1824), 2, 8, 24, 91, 98.

33 Pierre-François Kéraudren, *Mémoire sur les causes des maladies des marins, et sur les soins à prendre pour conserver leur santé dans les ports et à la mer, seconde édition* (Paris: Imprimerie Royale, 1824).

34 Numa Broc, "Les explorateurs français du XIXe siècle réconsidérés," *Revue française d'histoire d'outre-mer* 69, no. 256 (1982): 237–273.

35 See Étienne Taillemite, "Des marchands aussi et des stations navales," in idem, *Marins français à la découverte du monde*, 631–657, particularly 632–641.

36 Kéraudren, *Mémoire sur les causes des maladies des marins* (1824), 17–18.

37 Ibid., 34.

38 Ibid., 110–113.

39 Ibid., 58–59.

40 Ibid., 17–18, 30–31.

41 Ibid., 16, 18, 23, 40–48, 54.

42 Ibid., 12.

43 I discussed this episode in a previous study. See Daniel Dutra Coelho Braga, "Da Marinha francesa nos trópicos: apontamentos sobre o problema colonial da febre amarela nas Antilhas à época da Restauração Bourbon," *Dimensões – Revista de História da UFES* 41 (2018): 52–75.

44 François Xavier Donzelot, letter dated 19 February 1820, to the Baron Portal, SHD, Sub-series CC2, Codex 948.

45 Edme Michel Mauduit, letter dated 27 May 1820 to Pierre-François Kéraudren, SHD, Sub-series CC2, Codex 948.

46 François Xavier Donzelot, letter dated 19 February 1820, to the Baron Portal, SHD, Sub-series CC2, Codex 948.

47 See James McClellan III, *Colonialism and Science: Saint Domingue in the Old Regime* (Baltimore, MD: John Hopkins University Press, 1992).

48 Joseph-Fortuné-Theodore Eydoux, "Rapport au Conseil de santé de la marine à Toulon sur la campagne autour du monde de la corvette la Favorite, commandée par M. Laplace," certified copy, SHD, Sub-series BB4, Codex 1004.

49 Cyrille Laplace, letter dated 4 June 1832 to the Navy Minister, SHD, Sub-series BB4, Codex 1004.

50 Eydoux also wrote a report to the Navy health council of Brest, after having completed a circumnavigation with Nicolas Vaillant. See Joseph-Fortuné-Theodore Eydoux, "Rapport medical sur le voyage de circumnavigation exécuté par la Corvette de l'État la Bonite, dans les années 1836 et 1837, adressé au Conseil de Santé de la marine au port de Brest," SHD, Sub-series BB4, Codex 1007.

51 Kéraudren, *Mémoire sur les causes des maladies des marins* (1817), 2.

52 See Louis Marie Bajot, ed., *Annales maritimes et coloniales, IIe partie, vol. 1* (Paris: Imprimerie Royale, 1824), 457–541.

53 Pierre-François Kéraudren, "De la nourriture des équipages et de l'amélioration des salaisons dans la marine française," in Louis-Marie Bajot, ed., *Annales maritimes et coloniales, IIe Partie* (Paris: Imprimerie Royale, 1829), 362–381.

54 Ibid., 370.

55 On Albin Roussin, see Étienne Taillemite, *Dictionnaire des marins français: nouvelle édition revue et augmentée* (Paris: Tallandier, 2002), 469–470; Jean-Philippe Zanco,

"Roussin, Albin Reine, baron," in Jean-Philippe Zanco, dir, *Dictionnaire des Ministres de la Marine, 1689–1958* (Paris: SPM, 2011), 475–477.

56   Pierre-François Kéraudren, "De la fièvre jaune considérée par rapport à l'Europe et à la France; état actuel de la question," in Louis Bajot, ed., *Annales maritimes et coloniales, IIe Partie,* vol. 2 (Paris: Imprimerie Royale, 1824), 158–160.

57   We agree, therefore, with a point made by Michael A. Osborne, "Ressurrecting Hippocrates: Hygienic Sciences and the French Scientific Expeditions to Egypt, Morea and Algeria," in David Arnold, ed., *Warm Climates and Western Medicine: The Emergence of Tropical Medicine, 1500–1900* (Amsterdam/Atlanta, GA: Rodopi, 1996), 80–98, in particular, 94.

58   Michael Worboys, "Germs, Malaria and the Invention of Mansonian Tropical Medicine: From 'Diseases in the Tropics' to 'Tropical Diseases'," in David Arnold, ed., *Warm Climates and Western Medicine: The Emergence of Tropical Medicine, 1500–1900* (Amsterdam/Atlanta, GA: Rodopi, 1996), 198.

5 Positioning the North

Making British Geographical Knowledge of Australia in the Mid-Nineteenth Century

Johanna Skurnik

Introduction

In this essay, I investigate the elusive place of northern Australia in settler colonial imagination and colonial governmental practices. Present-day Northern Territory, an area of roughly 1,349,129 square kilometers, was originally the land of numerous Indigenous Australian peoples.[1] Under British occupation and invasion, it has since the nineteenth century undergone multiple different territorial designs and been the target of many colonization plans. The practices of governance and the diverse territorial ambitions portray ideas building on exploration knowledge, geographical deduction, and the spatial organization of colonial governance. The focus of this essay is the years of transformation from the late 1850s to the early 1860s, during which the position of the north as a site for planned colonization was reconceptualized. Prior to the early 1860s, the British made numerous attempts to establish settlements in the north, but these designs were consecutively abandoned.[2] The mid-century transformed the situation, at least in theory. As Dane Kennedy describes, this was the period for the "scramble for central and northern Australia," an era when the Australian colonial governments sought to extend their spheres of influence and to redraw the intercolonial boundary lines.[3] During this time, the colonies grew rapidly in the wake of the gold rushes and in the 1850s the non-Aboriginal populations of all colonies doubled. People were arriving at the colonies from Britain with the introduction of quicker and safer passages.[4]

The British explorations on the northern coast and the several attempts and final successes in traversing the continent from south to north and back made the north tangible in a novel manner. These included, among others, the North Australian Expedition led by Augustus C. Gregory in 1855–56 and the explorations by John McDouall Stuart around the turn of the 1860s. Gregory's expedition was the last major expedition funded by the British parliament and it was undertaken with high hopes regarding the geopolitical and economic opportunities that the north might offer.[5] Stuart conducted six expeditions between 1858 and 1862. Sponsored by land speculators and later by the South Australian government, Stuart explored central and North Australia and finally succeeded in crossing the continent in 1862.[6] Executed in the wake of the granting of "responsible government" for most of the Australian colonies, the geopolitical rivalries between the colonies affected how

DOI: 10.4324/9780367814540-6

government officials and the public interpreted their results.[7] The expeditions were part of a race that sought to expand settler colonialism to all sides of the continent, or as the contemporaries put it, to "complete" the occupation of the continent.

This chapter revisits these historical processes of exploration and territorialization to investigate the mobilities of geographical knowledge concerning the central and northern Australia and how different actors, especially civil servants, utilized it in their argumentation for the redrawing of the boundaries and the organization of effective colonial governance. An article printed in the Brisbane-based newspaper *The Courier* in 1861 aptly exemplifies the questions at stake. The newspaper printed the article after the government of Victoria had expressed interest to occupy a tract of land in the north. The colony based its claims on the discoveries made by the ill-fated explorers Robert O'Hara Burke and William Wills. The Victorian government had sent them to cross the continent but they both had demised whilst returning from their transcontinental expedition.[8] *The Courier* noted:

> It is manifestly absurd that a government [ie. Victoria] should demand the ownership and control of a territory separated from its own borders by country belonging to another state; yet such is the case in this instance, and it has occurred more than one in the history of Australian colonization. […] As every one properly versed in Australian geography knows, the territory lying between the 26th parallel of south latitude and the northern coast, and the 129th and 141st meridians of longitude, literally belongs to no colonial government at present. […] The claims preferred by the different colonies have, however, not been definitely decided upon by the home government, and the question therefore remains to be determined as to which province the coveted territory shall belong.[9]

These arguments put in circulation via the printing presses in Queensland underpin the focus of the present essay, namely the complex connections between land, geographical knowledge, and territorialization. They beg many questions: what was this "coveted territory" in the north? How did government officials, surveyors, explorers, and settlers know it to warrant it becoming the focus of heated geopolitical debate that generated piles and piles of letters and maps in different parts of the globe? Even though the area had been part of the mind map of many colonists in the south for decades, "northern Australia" was still in the 1850s and early 1860s a sparsely known area for the British. Furthermore, mapmakers territorialized it in many different ways, some labelling the north on their maps as "North Australia" as shown in Figure 5.1, although it was according to British legislation still part of New South Wales and such colonial territory no longer existed. Indeed, the British government had quickly abandoned the project of North Australia in the 1840s.[10]

Knowing the Australian north was a collective process, resulting from combinations of empirical observations, intercultural knowledge exchanges, speculations, and deductions in different locations. In this essay, I discuss the geography of these processes: adapting from David N. Livingstone and Charles W. J. Withers, knowledge's "some*where* is as vital as surveying and explaining its some*time* and

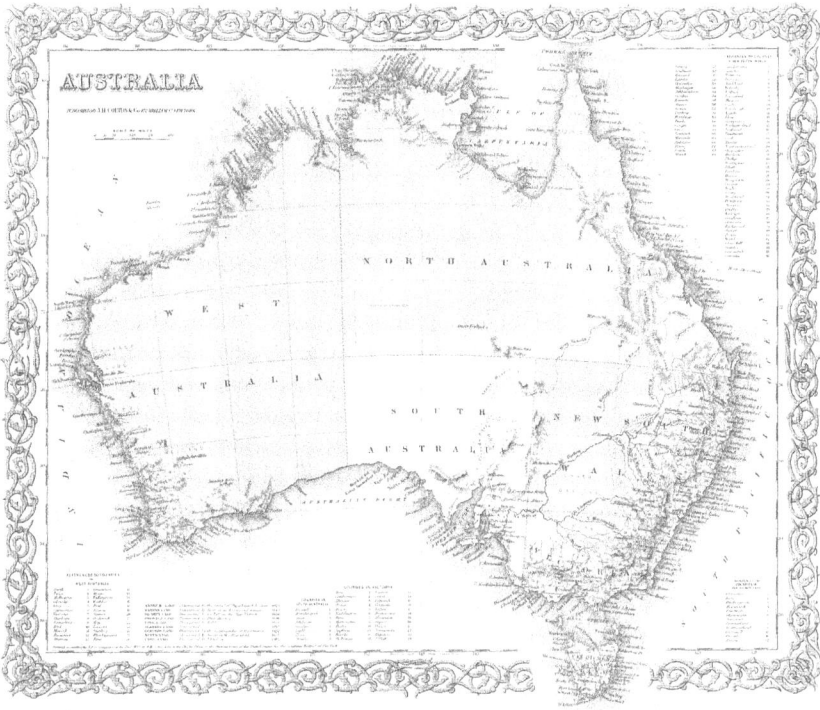

Figure 5.1 Map of Australia, 1855. Mapmakers visualized the territorial divisions of Australia in varied ways in the 1850s. This map by American mapmaker Joseph H. Colton, published in 1855, helpfully shows the positions of the then existing colonies Western Australia, South Australia, Victoria, and New South Wales. The map separates the majority of the north as "North Australia," although this colonial territory no longer existed. In addition, the western boundary of South Australia extends too far to the west.

Source: J. H. Colton, *Australia*, New York, NY: 1855, Wikimedia Commons.

some*bodies*." Thus, thinking geographically about knowledge helps understand how and why specific actions grounded by interpretations of knowledge emerged in different locations.[11] In political entities like the British Empire, spatiality permeated the practices of everyday knowledge-making and governance in a variety of ways that has been analyzed in "new imperial history" with the help of spatial metaphors such as webs, networks, and circuits. As Alan Lester notes, such approaches motivate examinations of "multiple meanings, projects, material practices, performances and experiences of colonial relations" that were shaped by the colonial connections between the colonies and the metropolis, each other and other localities.[12]

Recently, Kapil Raj has suggested a turn towards spaces of circulation in order to depart from what he sees as the limitations of networks, namely that they do not recognize power-relations and that the directions of knowledge flows cannot always be easily recognized. According to Raj, "spaces of circulation structure the way in which knowledge flows within closely knit social groups." They have social and physical geographies, can be discontinuous linking together separate spaces, and do not denote to a single European center. Consequently, Raj advocates a spatial concept that enables acknowledging the constraints inherent in a networked understanding of knowledge production.[13]

Applied in the context of this essay, Raj's concept offers a tool to analyze how government officials made geographies of colonial Australia. Geographical knowledge circulated in the Australian colonies and overseas via diverse media, both manuscript and print, composed and compiled by numerous agents. Additionally, an oral culture of communication linked actors together and sustained these circulations of knowledge. Uncovering their relative connections and disparities is key to understand the geographies of knowledge that existed. Such a task comes with methodological challenges that I tackle in this essay by approaching colonial and imperial archives as sites of knowledge production. Ann Laura Stoler argues in her study on the writerly forms of imperial governance that administrative documentation should be approached as "active, generative substances with histories, as documents with itineraries of their own" that reflect what she calls "colonial common sense." Scrutinizing the colonial archives helps dissect these processes of naming and producing ontologies as it highlights the epistemic anxieties that underpinned colonial governance: "grids of intelligibility were fashioned from uncertain knowledge; disquiet and anxieties registered the uncommon sense of events and things."[14] In this sense, the colonial archives evince the many ways that the "disturbance of distance" mattered in an ocean-spanning empire.[15]

Tony Ballantyne echoes Stoler when he stresses the need to understand the circulation of knowledge that generated colonial archives and their organization as well as critically examining their nodal position within what he calls the knowledge-producing webs of empire. Ballantyne conceptualizes imperial and colonial archives as hubs that testify for the centripetal processes that accumulate material to what becomes the archival collections. Simultaneously they "have a centrifugal function" as knowledge is distributed from them "through the act of reading, correspondence, the intertextual nature of print culture, or the exchange of manuscript or printed material."[16] Applied to the world of geographical discovery, then, the records of the colonial and imperial archives reveal the diverse people who sustained and connected the spaces of circulation such as explorers, indigenous peoples, mapmakers, settlers, and civil servants.

Taking these insights as starting points, I argue that the elusive place of northern Australia in colonial governance epitomizes the transformations of geographical knowledges into territorial governmentality. Based on manuscript and print material, this essay shows how different knowledge producers and material practices helped stich together knowledges that underpinned the territorialization of the north. The first section of the essay explores the epistolary networks and accumulation of colonial archives in different sites of knowledge production. Using Gregory's and Stuart's expeditions as examples, I investigate the routine practices that constituted the flows of knowledge and how this related to the conceptualization of the territory. I draw attention to the social and geographical factors that affected the circulation of knowledge. The second section investigates how the exploration knowledges transformed into practices of colonial governance. I analyze the visions of Queensland and South Australia that sought to make the continent legible in different ways. My analysis identifies the entanglements between the plans, reflects their position in the colonial and imperial mind maps, and analyses the eventual decisions that were made.

Geographical Knowledge in Transit

Viewed from the perspective of Colonial Office civil servants at Downing Street, London, the results of any expedition taking place in any part of the empire were intelligence to be handled in the appropriate manner. The execution of a government-funded expedition such as Gregory's involved piles and piles of paper, scrutinizing every single detail of the expedition, starting from its planning and ending with concerns over what should be done with the information and material that had been produced, and, importantly, how the knowledge should be applied in practice. From the 1830s onwards, the system of annotations and minutes filling up the back pages and corners of the dispatches offers glimpses into the empire's mundane practices of knowledge management that constituted colonial governance.[17] An intrinsic part of the daily tasks of the civil servants going through the bundles of dispatches and their enclosures which varied between maps, paintings, and specimens was to manage the information that the varied material communicated. This included – as the staff were not specialists of geography nor natural history – forwarding the material to those who were considered as such, and putting the materials received "in transit" yet again.[18] In the mid-nineteenth century, at the time of Gregory's and Stuart's expeditions, the Royal Geographical Society (RGS) and mapmaker John Arrowsmith functioned as the Colonial Office's closest allies, often complemented by the agents of the other metropolitan scientific institutions such as the Geological Society and the Royal Botanical Garden at Kew. For example, when information about the proceedings of Gregory's expedition started to flow to London either via Singapore or Sydney, the civil servants carefully determined who should receive what and with what kind of conditions. The RGS received copies of most of Gregory's expedition members' reports, and information concerning botany and the specimens collected were sent to Kew.[19] Arrowsmith gained access to the tracings.[20]

The system of transmission reflected the civil servants understanding of an order of knowledge in Britain but was also foregrounded by their ideas of publicizing information; Arrowsmith was expected to publish updated maps available to the public, and the scientific societies to discuss, examine, and display the material they received, thus determining their relative importance.

Publishing the expedition reports in Britain was a question of its own and demonstrates how the civil servants sought to make the results intelligible. Generally, most of the reports Australian explorers sent from the field to the settlements were published verbatim by the colonial newspapers. In Britain, the demand for accounts of Australian expeditions was much slimmer than for those concerning Africa.[21] However, being an expedition funded by the parliament, it was considered necessary to publish an account in Britain even though the expedition was not deemed interesting enough to warrant its lengthy printing.[22] Many options were considered to resolve what should be "done with these broken fragments of the result of N. Aust. Expedition" in order to publish a compact official account. For example, Charles Sturt, who had explored central Australia a decade before and advised in the planning of the North Australian Expedition, was consulted to provide a review of the diverse reports that had arrived. Eventually all these preparations proved unnecessary as Gregory's final report arrived and enclosed the particulars of the expedition in a condensed form. It was then printed with Sturt's report and a map by Arrowsmith.[23]

Similarly, allowing the RGS to publish Gregory's full journal was contemplated in reference to Gregory's assumed ambitions regarding the text. Most often, explorers themselves sought out rights from their sponsors to publish an account via a Britain-based book publisher, like John Murray or Thomas and William Boone.[24] In Gregory's case, this was considered unlikely. Clerk Gordon Gairdner summarized the situation in January 1858:

> We have no Official intimation on that subject, but I have been informed by the officers of the expedition with whom I have communicated that Mr. Gregory certainly has no such intention; and naturally so far there is no real variety in the descriptions of Australian exploration and Captain Sturt & Sir T. Mitchell have certainly worn out the subject. I was informed that Sir T. Mitchell's last publication was a losing speculation.[25]

Consequently, the RGS was allowed access to the journal with the condition that it would be returned immediately when the publication was complete as the journal was "strictly a record of this Department," and the full account of the proceedings of the expedition appeared in the *Journal of the Royal Geographical Society of London.*[26] These circuits of publishing reports mark instances of what Stoler calls the "pulse of the archive" and the "fine crafts of cribbing and culling."[27] What transpired provided links between the different spaces of circulation: leaps from the space of official correspondence and reporting into the public sphere as the travels on the Australian continent transformed into print in London.[28]

The civil servants' control over the documents relating to the North Australian Expedition contrasts with their inability to access the material relating to the first three expeditions of five conducted by John McDouall Stuart between 1858 and 1861. The expeditions' funders, land speculators James Chambers and William Finke controlled the information and material produced during these expeditions, with the South Australian government stepping in to fund the last two expeditions executed in 1860 and 1861. These coincided with the quest by the colony of Victoria to be the first to cross the continent and thus find a feasible route for the overland telegraph.[29] That Stuart's reports were not readily available for the governor

to consult and forward to London is visible, for example, in Richard MacDonnell's dispatch to London where he noted having gained access to only some of the
information:

> I have since had opportunities of reading Mr. Stuart's journal and of examin-
> ing his map of the new country, which he has found. I now enclose a tracing
> of the latter, which does not however give all the particulars noted in a very
> much fuller maps prepared by Mr. Stuart for the use of Mr. James Cham-
> bers – the gentleman who has hitherto depassed all the cost of Mr. Stuart's
> explorations.[30]

Consequently, the information that the governor of South Australia could forward
was limited by local interests which included by-passing the colonial government.
Indeed, the governor had received very general information which resulted in the
speculations of what Stuart had actually discovered, especially in relation to the
probable existence of an extensive gold field.[31] The situation escalated once Stuart
returned from his third expedition, again giving his report and tracing to his sponsor and friend Chambers. Even though MacDonnell had initially been promised
copies to help him report Stuart's findings to London, he did not receive them.
Instead, he could only forward newspaper articles in lieu of manuscript reports and
ended up having sealed documents at the government building, "on the condition
that the seals be not broken till the mail has left – an arrangement which seems
to afford satisfaction to Mr. Chambers."[32] MacDonnell reported the situation to
London:

> I can therefore only accompany this dispatch with a published extract from
> a letter of Mr. Stuart and also with an account of a banquet given to him and
> at which I had great pleasure in assisting – as the courage, energy and judge-
> ment of Mr. Stuart in accomplishing such an arduous undertaking with such
> inadequate means seem to be unrivalled in the History of Exploration on this
> Continent.[33]

This development in South Australia demonstrates one of the many factors that
related to the circulation of knowledge in one and its non-circulation in another,
connected social space. Chamber controlled the documents in order to protect his
own interests. Since the beginning of the 1850s, he, together with Finke, had been
interested to develop the northern hinterlands into a passage across the continent
and take advantage of the emerging lands.[34] Instead of passing the information via
the official correspondence, Chambers and Finke engaged directly with individuals linked with the London scientific societies. Their correspondents included men
like Roderick I. Murchison, then president of the RGS and known for his interest
in the colonization of the antipodes.[35] Private epistolary networks like these were
crucial vehicles through which the empire penetrated the daily lives of different
types of people, generating connections not only between metropolis and colony
but between colonies as well as beyond empire. The correspondence networks

and social relations in the colonies of land speculators like Chambers and Finke constituted a different space of circulation that connected with that of the civil servants only partially.[36] For a metropolitan organization such as the RGS and a scientist like Murchison, direct connections with settler colonists and explorers overseas helped generate the RGS into what Driver terms a center of information exchange.[37] Indeed, read at the Society's meeting in February 1861, the account of Stuart's explorations was noted to finally show "what Australia was made of."[38] At the Colonial Office, these chains of events adjusted the regular system of managing the circulation of material. The staff for example noted that Arrowsmith should be notified that the map was arriving via other route as per normal.[39] Eventually, however, Chambers granted permission for the government to print and disseminate Stuart's journal and lithograph his map which MacDonnell then speedily transmitted to London.[40]

The instances outlined above showcase how the flows of knowledge were as much processes of control as they were of transfer. Identifying these transits or their obstacles helps outline the discontinuous and connected spaces of circulation that came to exist in colonial Australia and Britain. Understanding the underlying circulations in cases like Gregory's and McDouall Stuart's expeditions helps comprehend the fluid nature and partiality of colonial archives and the temporalities of knowledge formation that the mobile documents helped generate. Caroline Cornish and Felix Driver highlight the importance of attending to the circulations that have constituted the creation of diverse institutional collections, in their case botanical collections.[41] I argue that in a similar manner as the distribution logs of the museum of economic botany, the Colonial Office correspondence with its minutes and annotations can be read as a record of the distribution of knowledge within and beyond government offices that led to the accumulation of different types of document collections. Simultaneously, they offer a starting point to analyze how the travelling knowledge received new meanings as it was received in different locations. The next section highlights this by analyzing how exploration knowledges transformed into ideas of territorial colonial governmentality.

Territorializing Geographical Knowledge

The British practices of territorializing the Australian continent developed from strategic interests, the realities of governance and land use and the mobilization of geographical knowledge to serve these arguments. A common discourse concerned the best way to divide the continent into productive and efficiently governed territories. This question had been debated for decades. For example, in 1838 the Royal Geographical Society published in its journal an article by Captain John Vetch discussing the best way to divide the continent into governable territories with similar resources. Building on the delineation of the recently established South Australian colony, Vetch suggested dividing the continent into nine territories roughly the size of the Iberian Peninsula (see Figure 5.2).[42]

Even though nothing came of Vetch's proposal at the time, the same discussions concerning how to best govern the vast continent by dividing it into territories and

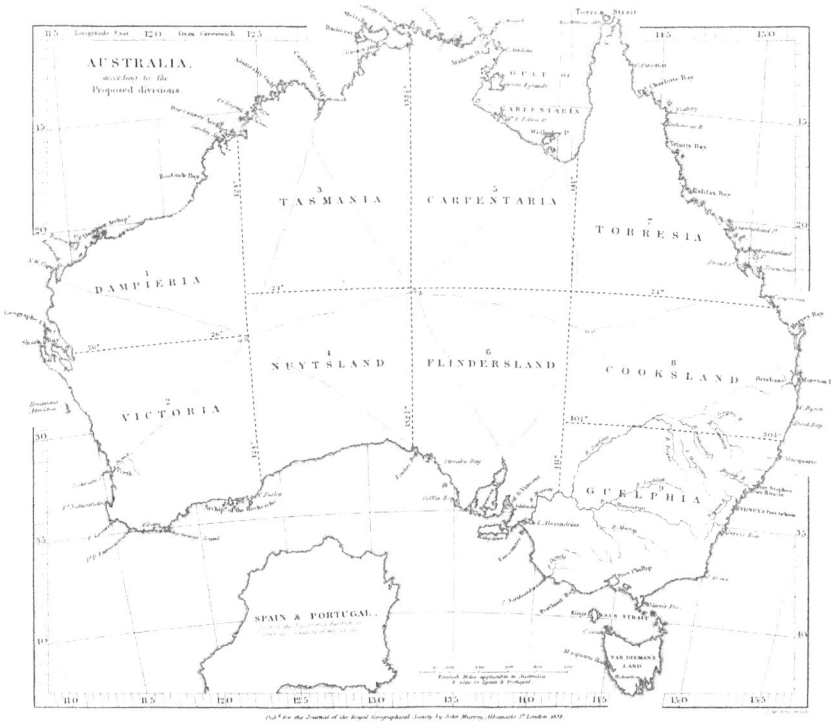

Figure 5.2 Map of Australia, 1838. John Vetch envisioned an Australia divided into territories that were approximately the size of the Iberian Peninsula, which he took to be a convenient size for a territory governed from one center. Mapmaker John Arrowsmith engraved the map, and the Royal Geographical Society published it accompanying Vetch's article in the Journal of the Royal Geographical Society, volume 8, 1838.

Source: University of Texas Libraries.

desires to completely occupy the continent emerged forcefully especially in the wake of Gregory's, Stuart's, and Burke's and Wills' expeditions in the beginning of the 1860s. In contrast to Vetch, however, the participants of these discussions could draw on diverse pieces of geographical knowledge deriving from the field. Exploration knowledge reported by the likes of Gregory and Stuart was used to support differing claims concerning the most natural and efficient way to redraw the boundary lines between the colonies. Following these arguments, this section analyses how the political geography of the continent was reconfigured in a particular space of circulation. This illustrates how the scramble of the continent connected with the circulation of geographical knowledge and its perceived links with the organization of colonial governance. When analyzed closely, these discussions showcase the meanings and consequences that geographical knowledge of northern Australia had for actors in different parts of the globe.

The British drawing and redrawing of territorial boundaries in Australia was constituted by their asserted sovereignty over the whole continent. Following James Cook's claim for the eastern half of the landmass, the British expanded their areas of influence by establishing other settlements, with that at Swan River enabling the claim for the whole continent. Unlike in other parts of the world, British invasion and dispossession of the indigenous peoples of the continent occurred without treaties.[43] It also transpired without serious rivalry from other European countries, making the continent a testing ground for the administration of vast tracts of land. At the most abstract level, the management of the continent's geography concerned the establishment of intercolonial boundary lines that defined the organization of governance and the extent of the governors' jurisdictions. Here maps helped enforce the connection between territory and the world "out there." In practice, knowing where the border line drawn according to a latitudinal or a longitudinal line lay on the ground was impossible and even when surveyed subject to dispute. It was also unnecessary when an area did not require direct administrative measures.[44] For decades, most of the boundaries of the colonies lay in what for the British was unmapped and unsettled country. This resembled what Lauren Benton has noted in the context of North America: "the specificity of geometric coordinates was set against indeterminacy," making the relation between imperial order and geographical information unstable.[45] Even though "natural" boundaries were the preferred means to design territories, in Australia these only surfaced on selected occasions like when the boundary of Victoria was assigned to the Murray River. Consequently, in the early 1860s the majority of boundaries were obscure, difficult to track down in the field, and anything but "natural." Furthermore, they did not in any way recognize the position of Indigenous Australian peoples or their territorialities and disabled their free use of land. Even though from 1860 the colonial governments started to set aside lands for Aboriginal people as reserves, the land remained in state control. The expansion to the north resulted in extended conflicts beyond governmental control.[46]

Pleas to reconsider how the north should be occupied originated both from London and the Australian colonies. Their shared destination was most often the Colonial Office and the secretary of state. For instance, in September 1860, after Stuart had just returned from his latest expedition during which he had reached the same latitudes from the south that Gregory had advanced to from the north, the governor of the newly founded colony of Queensland, George Bowen, sought for a clarification of the western boundary, of which confusion existed in the colony. Bowen argued that for the better administration of the continent the boundary should be extended to the 138° meridian to enable incorporating the so-called Plains of Promise and the "only safe harbour in the Gulf of Carpentaria" to the colony. Additionally, Bowen suggested establishing a new convict colony called Albert at the Victoria River, where convict labor could be beneficially used in isolation.[47]

Bowen supported his case by enclosing a memorandum and a map from Augustus C. Gregory, now surveyor-general of Queensland, and an extract from a text by John Dunmore Lang, "another writer of large Colonial experience," a colonist well known at the Colonial Office. Gregory bridged a connection with the physical geography of the country to the desirable location of the colony's boundaries as well as the shape of the new colony, for which a slice of Western Australia would be cut off to incorporate the small hills located by Gregory's party (see Figure 5.3). This suggestion significantly resembled those Gregory had submitted as part of his final report concerning the expedition a few years earlier and which had been printed in the parliamentary papers.[48] Similarly, Lang argued for the completion of the occupation of the continent by describing the possibilities that Victoria offered for the British once convict force had been used to lay the basis for a new settlement, or even a series of settlements.[49] A few months later, Bowen repeated his request and argued that "a glance at the Map will remind Your Grace that it [the area adjacent to the Gulf of Carpentaria] can be rendered practically available by Queensland alone" and the other colonies were separated from it, for example South Australia "by the whole breadth of the continent of New Holland."[50]

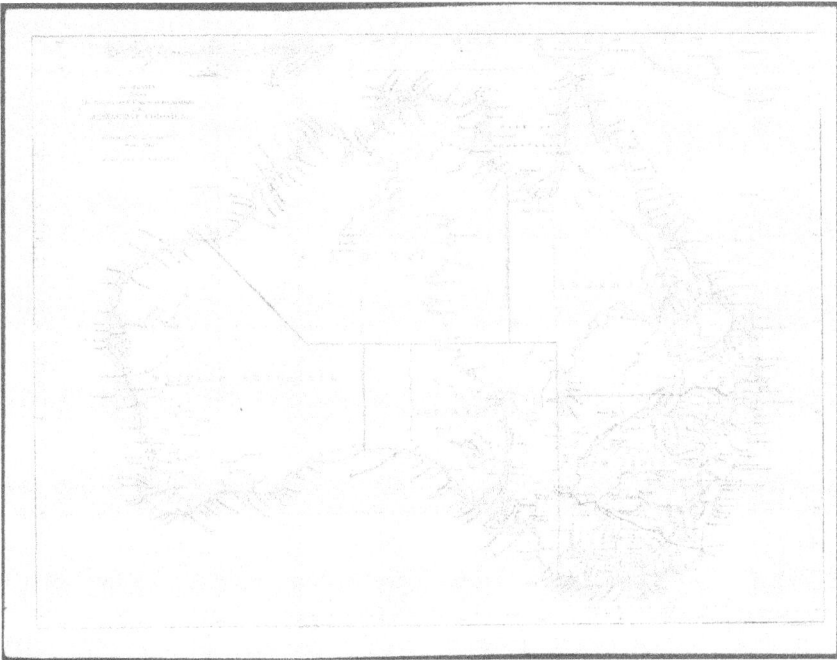

Figure 5.3 Map of Australia, 1860. Augustus C. Gregory's plan was to adjust the division of the northern parts of Australia into more governable territories. Governor of Queensland, George Bowen, sent the map, titled "Australia, showing the present and proposed boundaries of the respective colonies" to London to visualize Gregory's idea. The lines delineate the boundaries of the proposed colonies, with the suggested colony of Albert occupying the central north. The map also shows the position of the colony of Queensland which was separated from New South Wales in 1859.

Source: MAP RM 4459, National Library of Australia.

Bowen's dispatches effectively crafted an interpretation of the geography of the north and its implications for the practice of colonial governance. The suggestion concerning the use of convicts exemplifies how governable geographies were fashioned with the "material" provided by transportation. This seems to supplement Clare Anderson's argument that in the British Empire "the penal role of transportation" entangled with "its function as a means of labor supply, colonial governmentality and permanent settlement."[51] During this transformative period in convict transportation to Australia, the suggestion was troublesome. Indeed, following Gregory's expedition the matter had been considered, but consecutively abandoned.[52] The use of convict labor had constituted the functioning of the British settlements in New South Wales, Van Diemen's Land (present-day Tasmania), and Port Phillip until transportation had ceased to these destinations in 1840, 1849, and 1853, respectively. Starting in 1850, however, Western Australia, which had been formed without convict labor, had become a recipient of convicts and would continue to do so until 1868.[53]

In London, the civil servants were sympathetic of Bowen's need to locate the western boundary, considering that clerk Gordon Gairdner noted in his annotation that "I believe that there was no definite idea on the subject [when the colony was established]. The boundaries adopted were those furnished at last by the N. S. Wales Gov."[54] The suggestion for the creation of a new colony was not abandoned as such but it was deemed important not to entangle "Her Majesty's Government in any arrangement which may turn out inconvenient, on the ground of mere anticipations or projects, or on notions of fitness founded on our present geographical knowledge." Crucially, the reply letter to Bowen informed him that others were interested in the area too: the governor of South Australia, Richard McDonnell, had just send in an application for the extension of the South Australian territory to the north. This had occurred in the wake of the ill-fated success of the Victorian exploring expedition led by Robert O'Hara Burke and William Wills to cross the continent. The party, save one, had perished during their travel back to the south from the shore of the Gulf of Carpentaria. Furthermore, another group in Victoria had expressed desires to form a colony in the area.[55]

Consequently, a debate emerged over the most convenient and natural geographical connection to the area in the official correspondence, the colonial newspapers, and via other print products such as maps. The intercolonial connections via sea and post grew in the 1850s, and thus provided the media opportunities to increasingly report about the developments in the other colonies, such as the results of the expeditions, at a more rapid pace. The print media and the various pamphlets were important in mobilizing geographical knowledge in the public sphere.[56] The apparent availability of the north generated numerous privately organized designs for the colonization of the north immediately after Gregory's return and in the early 1860s. For example, one John Hall published in Melbourne, in 1862, a pamphlet defending the colonization of "Prince Albert Land" by establishing the first settlement, "Burke City," on the Albert River based on the latest exploration knowledge. Accompanied with a map, the pamphlet noted that "The colonization of the northern parts of this country, the most fertile of the Australian continent, has long been looked for. The time has now arrived."[57] Similar plans emerged also in London where for example in 1858 one P. MacAndrew from East London approached the secretary of state to suggest the establishment of a new post for cotton trade at the Gulf of Carpentaria.[58] These suggestions and ideas for the use of the northern lands showcase how geographical knowledges and designs (such as Gregory's Albert colony) concerning colonial Australia gained meaning and value in the social networks of diverse groups of people. They highlight how individuals with differing interests in making profit out of the Australian terrain eagerly picked up suggestions and speculations.

The argued expediency of establishing a new colony did not convince the Colonial Office. However, the smaller territorial claims of Queensland and South Australia were accepted and first in July 1861 the western boundary of South Australia and then in March 1862 the western boundary of Queensland were extended.[59] A turning point in Britain emerged from Charles Nicholson's letter to the secretary of state Duke of Newcastle urging the establishment of a new colony or the annexation of the territory to Queensland to organize the governance of the area.[60] Nicholson had returned to Britain from Australia in 1862, where he had for nearly thirty years careered, for example, as a physician, landowner, business developer, politician, and participated in the establishment of Australia's first university in Sydney.[61] He had close connections with the metropolitan scientists and had been immediately attached to the council of the RGS at the request of the Society's president Murchison.[62] The arrival of someone well versed with the developments of settlement and warning that it might become an area beyond government control (as the boundary Queensland was now officially fixed to the 138° meridian) appears to have concretized the need to divert from the "do nothing" policy Newcastle was known for. A temporary resolution was needed to generate tools to control the current and future squatters. A plan reflecting the metropolitan conceptualization of the easiest way to renew the political geography of the continent, drawing on the physical geography, was composed at the Colonial Office. The Duke of Newcastle forwarded to Queensland, New South Wales, and South Australia a suggestion where South Australia and Queensland both would acquire a portion of the northern area. The area below the tropics would be annexed to South Australia and that in the north to Queensland. Letters from Nicholson and the Colonial Land and Emigration Office, which the Colonial Office had consulted for advice, accompanied the dispatch.[63]

The civil servants learned the reactions from South Australia first: the governor Dominick Daly opposed the suggestion. Daly quoted John McKinlay, who had conducted the most recent explorations in the area and Daly argued that the area was most accessible from South Australia: "the migrations must come chiefly from the settled parts of the north of this colony (South Australia) or from the same latitudes of New South Wales, the Queensland frontier being barely practicable to cross stock from that Province with safety."[64] The report from the colony's executive council, attached to Daly's letter, enforced the idea of South Australia's natural connection to the area:

> [T]o annex the Victorian River territory provisionally to Queensland, would be to retard its occupation, while to attach it provisionally to South Australia,

would expedite its settlement: promote an extensive trade in horses with India, and bring these Colonies into more speedy telegraphic communication with Europe.[65]

For these South Australians, the north was a "new world" with countless economic possibilities, and they fashioned the movement in favor of its entire annexation as a uniform mission.[66] Not knowing what the opinion of Queensland would be, the civil servants reacted cautiously. Frederick Rogers noted that "I will only observe that the question is not who discovered the country nor (I should say) what would promote its most immediate colonization but by what Government it can be most effectively controlled." In his view, "it is much more likely to be connected with Brisbane of Q [...] But wait till we hear from Sir G. Bowen."[67]

To the Colonial Office's surprise, Bowen too opposed the metropolitan plan, but for other reasons. His dispatch stated that in fact the area would be best placed under the South Australian control. Bowen relied again on Gregory's recommendation about the natural way, following the known geography, to make the area governable. Gregory presented that a new colony placed under the control of an already existing government should be established. In his view, South Australia is the best candidate as Queensland did not currently have enough resources and Western Australia is not in a convenient or able position either. He stressed that this view countered the intuitive reading of a map of Australia. Gregory argued:

> [A] superficial glance at the map would perhaps tend to the Impression that Western Australia, within the nominal boundary of which a large extent of "Albert" is comprised, would be the first, Queensland the second, and South Australia the third colony in the list, as regards the facilities for local management.[68]

Queensland's opposition meant that only one option remained: South Australia. Thus, the Colonial Office ordered the Colonial Land and Emigration Office to immediately compose an order in council that would enable placing the territory under South Australian rule.[69]

Consequently, from July 1863 South Australia could boast having direct connection with the northern coast and the trade networks of the Indian Ocean. This lay the foundation for what would be called "the Great Central State" that quickly became a contested part of the colony's politics.[70] Eventually, South Australia's claims for the naturality of their control of the northern region, their capabilities in enforcing its governance, and most importantly their wishes regarding the available resources did not materialize as anticipated. Indeed, the administration and efforts to benefit from the economic potential resulted in a public debt of 4 million pounds. As early as 1866, the South Australian government reported to London about failures to locate suitable lands for settlement.[71]

According to Stoler, "partial understandings, epistemic confusion" were unavoidable parts of colonial governance. Tracking down instances of these helps uncover what she calls "competing conventions of credibility about what and whose evidence could be trusted."[72] To a certain extent, the decisions about northern

Australia concerned the credibility of the discussants. In Queensland, the opinion of Gregory, who was "universally admitted to be the man of all men best qualified to pronounce an opinion on this subject," heavily influenced the final position that the colony took regarding the central north.[73] In South Australia, reactions to news concerning the annexation of the territory were generally positive, and it is likely that its promoters' genuine belief in the areas' capabilities was a reflection of ignorance rather than deceit. Some doubts did start to emerge when due to a political crisis a new government was formed with members who had disputed the annexation. Consequently, in the coming years the area "had to be settled by men […] who had opposed its acquisition as a waste of public money."[74]

Conclusion

This essay has analyzed the reconfiguration of the political geography of Australia in the wake of the explorations of the 1850s and early 1860s. The "scramble for Australia" and the shifting position of the scarcely known central north exemplifies how geographical knowledge was utilized as proof for competing purposes by individuals and civil servants of the colonial and imperial governments. Knowledge as well as ignorance constituted the successful argumentations for the most convenient way of organizing the governance of the vast continental space and creating governable territories. Tony Ballantyne has noted that "nineteenth-century colonial culture was by its very nature porous, fluid and […] it was energised by the circulation of people, ideas, and ideologies through and almost bewildering array of institutions, networks, and forms of cultural production."[75] The colonial archives analyzed in this chapter highlight this porousness but simultaneously offer clues to understand the processes that made knowledge and pursued action. Investigating the territorialization of northern Australia highlights the "inter-regional and transnational structures that enabled the integrative, if highly uneven, work of modern imperial systems."[76] Following the movements of material like the reports of explorers in transoceanic and colonial spaces reveals how different spaces linked together. The colonial archives reflect the spaces of circulation they were part of both physically and socially. Approached as sites of knowledge production, these archives like the Colonial Office's, therefore, demonstrate the geographies of making geographies of colonial Australia.

Notes

1 The Northern Territory is presently one of the most diverse linguistic areas in the world. See David H. Horton, *The Aiatsis Map of Indigenous Australia* (Australian Institute of Aboriginal and Torres Strait Islander Studies, 1996) for reference of the diversity of the language, social, and nation groups of Indigenous Australia.

2 Peter Donovan, *A Land Full of Possibilities. A History of South Australia's Northern Territory* (St. Lucia, London, New York: University of Queensland Press, 1981), 10–20; James M. R. Cameron, "The Northern Settlements: Outposts of Empire," in Pamela Statham, ed., *The Origins of Australia's Capital Cities* (Cambridge: Cambridge University Press, 1989).

3 Dane Kennedy, *The Last Blank Spaces. Exploring Africa and Australia* (Cambridge: Harvard University Press, 2013), 104.

4 Stuart MacIntyre and Sean Scalmer, "Colonial States and Civil Society, 1860–90," in Alison Bashford and Stuart MacIntyre, eds., *The Cambridge History of Australia. Volume 1. Indigenous and Colonial Australia* (Cambridge: Cambridge University Press, 2013), 189, 194.

5 Norman Etherington, "Recovering the Imperial Context of the Exploration of Northern Australia," in Anne M. Scott, Alfred Hiatt, Claire McIlroy, and Christopher Wortham, eds., *European Perceptions of Terra Australis* (Farnham: Ashgate, 2011), 233–246.

6 Deirdre Morris, "Stuart, John McDouall (1815–1866)," *Australian Dictionary of Biography*, available at: http://adb.anu.edu.au/biography/stuart-john-mcdouall-4662/text7707 (accessed 1 January 2020).

7 Responsible government was granted to the settler constituencies in the following order: Victoria (1851), New South Wales (1855), South Australia (1856), Queensland (1859), and Western Australia (1890).

8 For the expedition, see Sarah Murgatroyd, *The Dig Tree: The Story of Burke and Wills* (London: Bloomsbury, 2002).

9 *The Courier,* 18 December 1861.

10 Johanna Skurnik, *Cartographic Histories of the Western Territorialization of Northern Australia, 1840s–1900s. Global Circuits of Knowledge and the Mapmakers' Craft* (Munich: De Gruyter Oldenbourg 2020), 331–358.

11 David N. Livingstone and Charles W. J. Withers, "Thinking Geographically About Nineteenth-Century Science," in *Geographies of Nineteenth-Century Science* (Chicago, IL: University of Chicago Press, 2011), 3.

12 Alan Lester, "Imperial Circuits and Networks: Geographies of the British Empire," *History Compass* 4, no. 1 (2006): 131–132. Also see Zoë Laidlaw, *Colonial Connections, 1815–45: Patronage, the Information Revolution and Colonial Government* (Manchester: Manchester University Press, 2005); Brett M. Bennett and Joseph M. Hodge, eds., *Science and Empire. Knowledge and Networks of Science across the British Empire, 1800–1970* (Palgrave Macmillan, 2011).

13 Kapil Raj, "Networks of Knowledge, or Spaces of Circulation? The Birth of British Cartography in Colonial South Asia in the Late Eighteenth Century," *Global Intellectual History* 1883 (2017), 1–18; idem, "Spaces of Circulation and Empires of Knowledge. Ethnolinguistics and Cartography in Early Colonial India," in Paula Findlen, ed., *Empires of Knowledge. Scientific Networks in the Early Modern World* (London: Routledge, 2018), 275, 287–288.

14 Ann Laura Stoler, *Along the Archival Grain. Epistemic Anxieties and Colonial Common Sense* (Princeton, NJ: Princeton University Press, 2009), 1–3.

15 I am quoting from Felix Driver, "Distance and Disturbance: Travel, Exploration and Knowledge in the Nineteenth Century," *Transactions of the Royal Historical Society* 14 (2004), 73–92.

16 Tony Ballantyne, *Webs of Empire: Locating New Zealand's Colonial Past* (Vancouver: UBC Press, 2014), 180–181, 190–191.

17 Their introduction related to the renewal of administrative practices, see Laidlaw, *Colonial Connections,* 49–54.

18 Cf. James A. Secord, "Knowledge in Transit," *Isis* 95, no. 4 (2004): 654–672.

19 For example: Annotations in Denison to Labouchere 20 March 1856 no 52, CO201/493, The National Archives (hereafter TNA); Annotations and draft letters in Foreign Office to Merivale 4 November 1856, CO201/496, TNA; Merivale to Hooker 19 June 1857 and Merivale to Hooker 24 September 1857, Miscellaneous reports, MCR/7/5/10, Royal Botanical Gardens.

20 Minute by Gairdner in RGS to Colonial Office 29 October 1856, CO201/496, TNA.

21 Kennedy, *The Last Blank Spaces*, 254.

22 Annotations in Denison to Labouchere 4 April 1857 no 65, CO201/498, TNA.

23 Annotations in Denison to Labouchere 22 June 1857 no 93, CO201/499, TNA; *Papers relating to an expedition recently undertaken for the purpose of exploring the northern portion of Australia,* Presented to both Houses of Parliament by command of Her Majesty, London 1857.

24 Kennedy, *The Last Blank Spaces,* 254. For the practices of publishing see Innes M. Keighren, Charles W. J. Withers, and Bill Bell, *Travels into Print: Exploration, Writing, and Publishing with John Murray, 1773–1859* (Chicago, IL: University of Chicago Press, 2015).

25 Minute by Gairdner 21 January in Murchison to Colonial Office 19 January 1858, CO201/506, TNA.

26 Minute by Gairdner 21 January in Murchison to Colonial Office 19 January 1858, CO201/506, TNA. See *The Journal of the Royal Geographical Society of London,* 28 (1858): 1–137.

27 Stoler, *Along the Archival Grain,* 19–20.

28 Cf. Keighren, Withers, and Bell, *Travels into Print.*

29 Kennedy, *The Last Blank Spaces,* 104–105.

30 MacDonnell to Lytton 14 August 1859 no 338, CO13/100, TNA.

31 MacDonnell to Lytton 14 August 1859 no 338, CO13/100, TNA.

32 MacDonnell to Newcastle 26 October 1860 no 436, CO13/102, TNA.

33 MacDonnell to Newcastle 26 Oct 1860 No 436, CO13/102, TNA.

34 Jack Cross, *The Great Central State. The Foundation of the Northern Territory* (Kent Town: Wakefield Press, 2011), 4–6.

35 Annotations in MacDonnell to Newcastle 26 October 1860 no 436, CO13/102, TNA; *Proceedings of the Royal Geographical Society of London (PRGS)* 5, no. 2 (1861): 55–60; *PRGS* 5, no. 3 (1861): 104–106. For Murchison see Robert A. Stafford, *Scientist of Empire: Sir Roderick Murchison, Scientific Exploration and Victorian Imperialism* (Cambridge: Cambridge University Press, 1989), chapter 2.

36 See Ballantyne, *Webs of Empire*; Laura Ishiguro, *Nothing to Write Home about: British Family Correspondence and the Settler Colonial Everyday in British Columbia* (Vancouver: UBC Press, 2019).

37 Felix Driver, *Geography Militant: Cultures of Exploration and Empire* (Oxford: Blackwell Publishers, 2001), 36–37.

38 *Adelaide Observer,* 27 April 1861.

39 Annotation by Gairdner in MacDonnell to Newcastle 26 October 1860 no 436, CO13/102, TNA. Compare annotations in MacDonnell to Newcastle 26 December 1860 no 469, CO13/103, TNA.

40 MacDonnell to Newcastle 26 December 1860 no 469, CO13/103, TNA.

41 Caroline Cornish and Felix Driver, "'Specimens Distributed': The Circulation of Objects from Kew's Museum of Economic Botany, 1847–1914," *Journal of the History of Collections* 32, no. 2(2020): 327–340.

42 Considerations on the Political Geography and Geographical Nomenclature of Australia, *Journal of the Royal Geographical Society* 8 (1838): 157–169.

43 Saliha Belmessous, "The Tradition of Treaty Making in Australian History," in Saliha Belmessous, ed., *Empire by Treaty: Negotiating European Expansion, 1600–1900* (Oxford: Oxford University Press, 2014), 186–213. For the early history of the newcomers, see Shino Konishi and Maria Nugent, "Newcomers, c. 1600–1800," in Alison Bashford and Stuart MacIntyre, eds., *The Cambridge History of Australia. Volume 1. Indigenous and Colonial Australia* (Cambridge: Cambridge University Press, 2013), 43–67; Grace Karskens, "The Early Colonial Presence, 1788–1822," in Alison Bashford and Stuart MacIntyre, eds., *The Cambridge History of Australia. Volume 1. Indigenous and Colonial Australia* (Cambridge: Cambridge University Press, 2013), 91–120; Lisa Ford and David Andrew Roberts, "Expansion, 1820–1850," in Alison Bashford and Stuart MacIntyre, eds., *The Cambridge History of Australia. Volume 1. Indigenous and Colonial Australia* (Cambridge: Cambridge University Press, 2013), 121–148.

44 An illustrating example in the Australian context is the boundary between South Australia and New South Wales (and later Victoria) which was debated and surveyed several times from the 1830s on. See Gerard Carney, "A Legal and Historical Overview of the Land Borders of the Australian States," *Australian Law Journal* 90, no. 8 (2016), 590–593.

45 Lauren Benton, *A Search for Sovereignty: Law and Geography in European Empires, 1400–1900* (Cambridge: Cambridge University Press, 2010), 14.

46 MacIntyre and Scalmer, "Colonial States and Civil Society, 1860–90," 214–215.

47 Bowen to Newcastle 30 September 1860, no 79, CO234/2, TNA. The confusion had emerged from the publication of the NSW Law Officer's opinion regarding the position of the boundary as one following the 141st meridian.

48 Compare Enclosure 1 in Bowen to Newcastle 30 September 1860, no 79, CO234/2, TNA and *Papers relating to an expedition recently undertaken for the purpose of exploring the northern portion of Australia.* The map was John Arrowsmith (lith.), *Australia, showing the present and proposed boundaries of the respective colonies. Enclosure no. 2 in Dispatch no. 79, 30 Sept. 1860, Sir G.E. Bowen to Secretary of State* (London, 1861), MAP RM 4459, National Library of Australia (hereafter NLA).

49 Enclosure 3 in Bowen to Newcastle 30 September 1860, no 79, CO234/2, TNA. I have not been able to identify the source of the extract.

50 Bowen to Newcastle 8 December 1860 no 92, CO 234/2, TNA.

51 Clare Anderson, "Transnational Histories of Penal Transportation: Punishment, Labour and Governance in the British Imperial World, 1788–1939," *Australian Historical Studies* 47, no. 3 (September 2016): 396.

52 See annotations in Chimmo to Colonial Office 20 January 1857, CO201/501, TNA; Minute in Bowen to Newcastle 30 September 1860, no 79, CO234/2, TNA.

53 Anderson, "Transnational Histories of Penal Transportation, 1788–1939," 382–383.

54 Annotation by Gairdner in Bowen to Newcastle 30 September 1860, no 79, CO234/2, TNA.

55 Newcastle to Bowen 26 February 1861, published in *The Courier* 20 May 1861.

56 MacIntyre and Scalmer, "Colonial States and Civil Society, 1860–90," 194.

57 10491.bbb.38.(6.),7, British Library (BL). Albert Land was also mentioned in an atlas of Australia published in Melbourne in 1863 by French-born Frederick Proeschel (1809–1870). See Maps 32.D.37, BL.

58 MacAndrew to the Secretary of State 23 April 1858, CO201/507, TNA.

59 Carney, "A Legal and Historical Overview of the Land Borders of the Australian States," 590, 593.

60 House of Commons papers, no 506, North Australia. Copy or extracts of correspondence between the Secretary of State for the Colonies and the governors of the Australian colonies, respecting the annexation of the crown lands around the Gulf of Carpentaria, 17–19.

61 David S. Macmillan, "Nicholson, Sir Charles (1808–1903)," *Australian Dictionary of Biography*, available at: http://adb.anu.edu.au/biography/nicholson-sir-charles-2508/text3387 (accessed 1 January 2020).

62 Roderick Weir Home, *Australian Science in the Making* (Cambridge: Cambridge University Press, 1988), 90.

63 See draft letters to governors and Nicholson in Murdoch to Rogers 19 August 1862, CO234/7, TNA; Cross, *The Great Central State*, 10–11.

64 Daly to Newcastle 26 November 1862 no 68, CO13/110, TNA.

65 Enclosure 1 in Daly to Newcastle 26 November 1862 no 68, CO13/110, TNA. These points were repeated in the next dispatch where the results of Stuart's successful overland expedition were reported. See Daly to Newcastle 26 December 1862 no 73, CO13/110, TNA.

66 Donovan, *A Land Full of Possibilities*.

67 Annotation by Rogers in Daly to Newcastle 26 November 1862 no 68, CO13/110, TNA.

68 Enclosure no 2 and Draft letter to Colonial Land and Emigration Commissioners in Bowen to Newcastle 18 January 1863 no 5, CO234/8, TNA.

69 Enclosure no 2 and Draft letter to Colonial Land and Emigration Commissioners in Bowen to Newcastle 18 January 1863 no 5, CO234/8, TNA.

70 Cross, *The Great Central State*, 15–16.

71 Daly to Carnavaron 27 September 1866 no 45, CO13/119, TNA; Donovan, *A Land Full of Possibilities*, xx.

72 Stoler, *Along the Archival Grain*, 185.

73 Bowen to Newcastle 18 January 1863 no 5, CO234/8, TNA.

74 Donovan, *A Land Full of Possibilities*, 38–40; Cross, *The Great Central State*, 7, 17–18.

75 Ballantyne, *Webs of Empire*, 191.

76 Ibid., 193.

6 Maps and the Man on the Spot

Bio-geographies, Knowledge, and Authority around and about the Zambezi

Elizabeth Haines

Introduction: Maps and the Man on the Spot

In August 1900, an agent of the British South Africa Company (BSA) set up camp by the Kafue River, a large tributary of the Zambezi. This agent, Valdemar Gielgud, was an American, a veteran of military action by the BSA further south, and had held two terms of office as a local administrator for the company, from the first of which he was discharged as unfit.[1] His correspondence reveals an angry and aggressive individual, whose attitude made even the thick-skinned directors of the BSA a little uncomfortable. His task in the Hook of Kafue was to enforce a geographical extension of BSA rule, which he carried out over a period of two years, sending reports back to his superiors locally and in London, accompanied by maps that charted his activity:

> This patrol passed through much new country and visited many Kraals and much work was done […] I enclose a list of Chiefs visited. Those visited for the first time are marked with a star. I also enclose a new map of the Hook of the Kafue, showing all chiefs who have been visited and made submission and who visit at my camp.[2]

Through Gielgud's reports, his journeying travelled beyond the Hook, back to London, which made the physical and human geography of the region legible. Through this process, a British colonial territory in the region was constructed, a geographical entity that eventually became known as Northern Rhodesia (then at independence, Zambia). Gielgud's maps have been read as founding documents in the creation of that colonial state, keystones in the installation of a vertical state architecture that eventually matched geographies of administration on the ground.[3]

That reading sits in line with much scholarship on imperial governance that has emphasized the importance of centralized knowledge in defining and controlling imperial territory. In the ideal form of this model (that I will call "centrist"), knowledge travelled to the center of power, or center of calculation, and reciprocally instructions emanated *from* that center. In the centrist model, the geographies of knowledge and the geographies of power were co-extensive and symmetrical. Despite substantive and important critiques of this model for understanding imperial

DOI: 10.4324/9780367814540-7

power, particularly from anthropological studies of governance, and global histories of knowledge, the effect of this framework on scholarship remains significant.[4] Centrist modes of understanding colonial authority have been particularly influential in studies of the role of mapping in European empires, as a mode of visually and ontologically ordering colonial spaces and rendering them legible to the center. This model for understanding the relationship between mapping and colonial authority has been persistent, even where historians have demonstrated flawed or failed attempts to achieve that legibility.[5]

It has, however, been difficult to reconcile the "centrist" model for the relationship between knowledge and authority with historical political geographies of British colonial rule in sub-Saharan Africa for several reasons. One primary reason for this was the British reliance on indirect rule. Enrolling Africans into the British system of governance as chiefs and headmen created a relationship of both political and epistemic dependence on African authority.[6] Despite the fact that colonial authorities regulated the power of African authorities by supervising their budgets, legal rulings, and economic leadership, this dependence created a level of opacity in how central government might "see" the territory.

Secondly, studies of British colonial administration in Africa have shown the systems of reporting and bureaucracy to be relatively weak.[7] Colonial officers were not primarily bureaucrats, sending knowledge upwards in the governmental hierarchy. The foremost role of administrators was to represent British government *in* the colonized territories "an impressive, even awesome and omnipotent embodiment" of imperial power.[8] As the Resident Commissioner of North-Western Rhodesia once put it, local administrators were supposed to be "the centre of influence in every district."[9] From this perspective on colonial authority (that I will call "de-centrist"), the cult of personal authority created a political structure in which imperial power was centrifugal.[10] Such accounts present geographies of power that seem much more credible than centrist models of authority; however, in focusing on personal authority, they often ignore or flatten the role of knowledge-making practices. Knowledge is often, in these paradigms, expressed as an absence. For the political scientist Jeffrey Herbst, this system produced a situation in which much of colonial administration was improvised by the "man on the spot."[11]

In order to resolve the gap between the centrist and de-centrist accounts of governmental authority, between legible and illegible forms of governance, we need to consider the geographies of power and knowledge as dynamic and multi-directional. Recent literature in the mobility "turn" has ignited closer interest in long-distance routes and African social networks that had pre-colonial forms and persisted through and beyond colonial rule.[12] Exerting power through this kind of socio-political system relied on personal mobility, relays, exchanges, and translation. Some literature has explored the resistance of these practices to colonial intervention.[13] However, we have yet to fully understand the co-option of these practices by colonial authority. This chapter draws out some examples.

In other scholarship on imperial governance, power has been considered within "process geographies."[14] From this perspective, mobility was and is a creative and productive political instrument for powerful authorities in and of itself.

For example, Benton describes how imperial control spread through travel and negotiation. She points to patterns of political control that were "variegated" and manifested in corridors and enclaves: "tubulous" sovereignty.[15] Schaffer et al. emphasize the role of intermediaries in what they call "government by go-betweens."[16] Yet this scholarship has not been brought into direct dialogue with what we understand about the role of documents in making colonial states. Studies of the intersection between mobility and administrative practices of European colonial powers have so far remained focused on areas considered as border or frontier spaces, or that have framed mobility as a tool for "proto-administrative" practices, a precursor to proper forms of government.[17] This chapter indicates how, on the contrary, mobility was and remained at the heart of British colonial administrative structures right up to Zambian independence in 1964. In sum, we need better ways to reconcile mobility, bureaucracy, and colonial territoriality.

In search of that reconciliation, in this chapter, I translate Paul Carter's impetus to "recover the movement history that underwrites geography" to the history of the construction of the colonial state.[18] Carter's approach allows us to build richer understandings of the role of circulating agents and documents in the production of colonial territory. Here I use the principle of "movement histories" to explore the careers of two imperial agents, Colin Harding and Valdemar Gielgud, who were at work in the Zambezi region in the early twentieth century. Their endeavors to exert colonial authority deployed multiple forms of political practice that were highly mobile, volatile, and situated, including both diplomacy and violent military engagement. Their individual movement histories offer us an opportunity to take a closer look at how mobility created a tubulous sovereignty that rehearsed existing African social geographies. They also offer us a means to consider alternative roles for geographical knowledge in the process of building a colonial government in the region: not only focusing on what was communicated and passed on, but also the colonial state's "self-willed amnesia."[19]

Both agents contributed content for new imperial maps. The maps they drew up as accounts of their journeys were at first read at face value: as the articulation of a colonial territory produced by harnessing pre-existing routes and socio-political networks. Yet in London, the maps were then compiled with other cartographic sources. The resulting compilations, drawing on typical cartographic conventions, presented the emergent territories ahistorically, and as apparently politically homogeneous territories. In these new visions of the Zambezi region, the individuality of the officers' accounts and their motion in the field were gradually erased. Maps that transcribed the messy and variegated nature of sovereignty in the region and that showcased the authors' bio-geographies[20] were replaced with images of orderly and uniform zones of British jurisdiction, "flat and smooth, equally authoritative in every part."[21]

Despite this smoothing of cartographic representation, if we read "along the grain" of that apparent transition in cartographic form, we see that the tensions between territory, journey, bio-geography, and authority that are so evident in Harding and Gielgud's careers were not resolved.[22] Those tensions instead produced a divergence in cartography: two different sets of colonial visions of Northern

Rhodesia that were subsequently held in uncomfortable parallel until the end of British occupation. The differing patterns of circulation of these two genres, the ways in which maps and documents travelled (or did not travel), indicate the ways in which London was able to *liberate* itself from particular kinds of knowledge. The set of practices through which the territory came to be inscribed into archives and discourses at the Imperial center as a "flat and smooth" political space were not only techniques for the accumulation of knowledge. This flattening and smoothing was a procedure for the colonial government to forget pre-colonial political and social geographies in the territory as Carter has suggested.[23] However, in the case of Northern Rhodesia, it was at least as much a procedure for the colonial authorities to occlude the mechanics of the political modality from which they themselves derived their power.

Routes and Spheres of Influence in the Zambezi Region of the Early Twentieth Century

Colin Harding and Valdemar Gielgud were veterans of pacification battles that had been fought in Southern Rhodesia, and at the close of the nineteenth century both were posted to the Zambezi basin to forge the relationships that would define yet further British territory. Harding's role centered on the British relationship with the Lozi elite, based in Lealui (Upper Zambezi). Gielgud was posted to a new camp "Mwenga River" in an area between the Kafue and the Lower Zambezi, known as the Hook of Kafue. The Hook of Kafue lay on the north-eastern fringes of Lozi authority, on the south-western boundary of a territory that the BSA had been expanding from the north, which in 1900 was formerly claimed and titled North-Eastern Rhodesia (see Figure 6.1). The relationship between the new colonial territories of North-Western Rhodesia and North-Eastern Rhodesia, and their relationship to existing African polities and power structures were being defined.

In order to better understand the relationship between the circulation of people and the circulation of documents, I am going to look more closely at the *practices* deployed by Harding and Gielgud as administrators, and particularly their mobility. We will see that their authority was produced by journeying that relied on existing social and political geographies in the territory, routes, networks, and political modalities – both African and European – that pre-dated colonial administration. In Harding's case, access to these routes and networks was more often achieved by political diplomacy with existing African authorities. In Gielgud's case, access was achieved predominantly by violence and coercion. In both Harding and Gielgud's cases, colonial influence was brought to bear through embodied and mobile manifestations of British political strength.

Harding's Mobility

Colin Harding (1863–1939) was the son of impoverished gentleman farmer. He arrived in Bulawayo in 1894 and began work as a builder's laborer, then as a clerk, before dealing in gold claims.[24] In 1896, Harding signed up as a soldier to

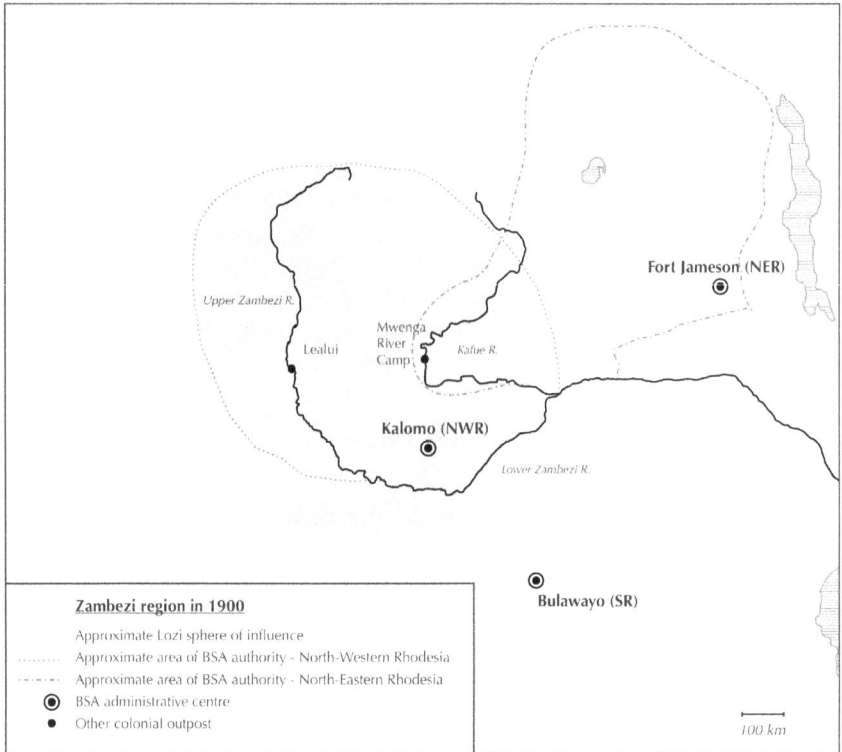

Figure 6.1 The Zambezi region in 1900. The figure illustrates some of the sites and zones of colonial authority in the first decades of the twentieth century, a moment of change in the region. In the last decades of the nineteenth century, through the warring and politicking of Cecil Rhodes, the British South Africa Company (BSA) had pushed its activities northward towards Central Africa. In 1890, negotiators encouraged by Rhodes and ratified by the crown persuaded the king of a large pre-colonial polity, the Lozi, to sign a treaty under which the British offered protection. As a result, the Lozi kingdom (centered around the Upper Zambezi) became part of BSA chartered territory. Figure made by the author.

participate in the First and Second Matabele Wars and rose to Major by 1897.[25] After a period of leave in England in 1899, he was appointed to assemble and train a new force of "native police." That police force was to be active in the newly formed Protectorate "Barotseland," an African polity that was providing fragile legitimacy to British territorial claims in the Zambezi region.

Harding's publications, and his correspondence, allow us to chart his activities around the Zambezi very closely.[26] The Bodleian Library holds a diary by Harding's brother and companion, William, which includes rough pen and pencil sketch maps as a form of route note-taking in the field (Figure 6.2).[27] The combination of these documents allows us to see that Harding's career in the Zambezi region was characterized by constant movement. On appointment as commander of the

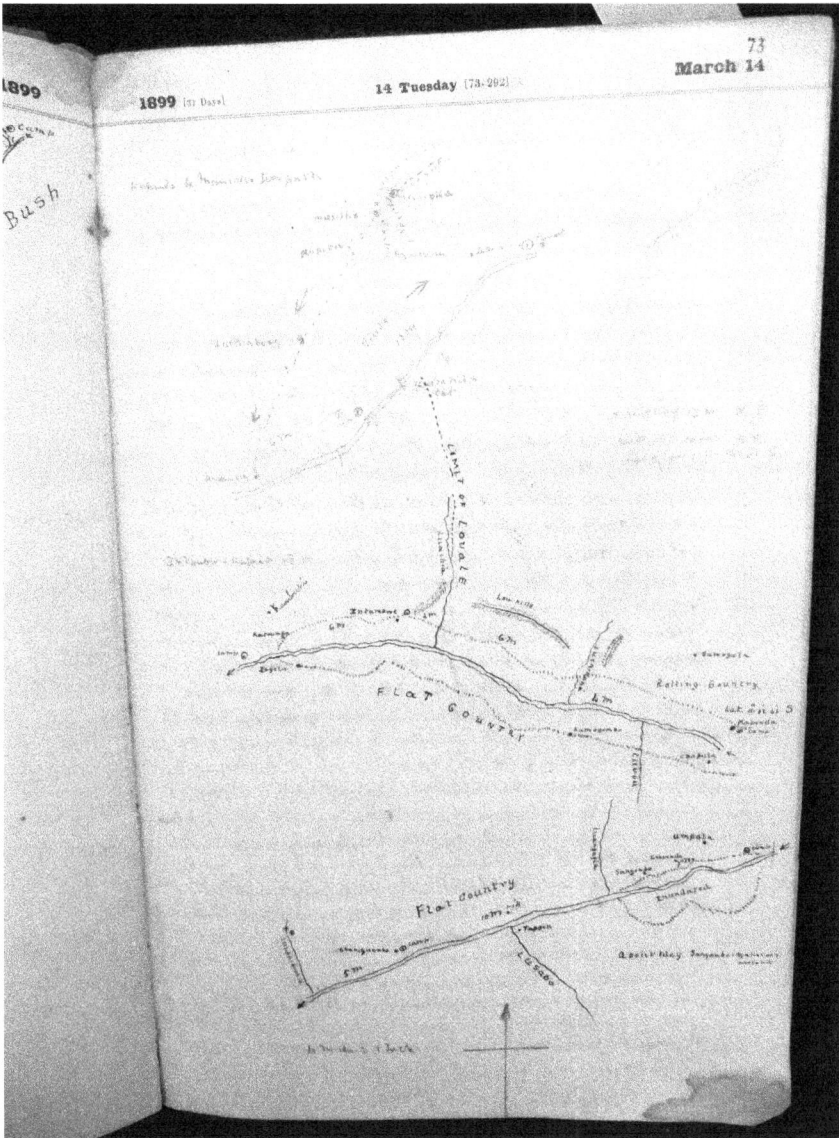

Figure 6.2 A series of sketch maps from William Harding's diary. The figure shows the routes taken by Harding and his brother Colin in journeys along the Zambezi to the North of Lealui during one episode of Harding's extensive travelling between 1899 and 1901.

Source: Bodleian Library MS. 5433.

Barotse Native Police, he started out from Bulawayo, in Southern Rhodesia, for Lealui, the capital of the Lozi kingdom in September 1899. Within the first 14 days of his new employment, he had already travelled 450 miles, by taking an extra trip to inspect the stations at Kalomo and Monze whilst he waited for his escort to catch up with him. In less than a year, Harding travelled to the source of the Zambezi, to the source of the Kabompo river, in what is now Angola, as well as journeying West towards the Kafue, the Batoka Plateau, and the Lower Zambezi valley. Wright has calculated that in less than a year Harding travelled 2,235 miles back and forth across the region.[28]

Harding's constant circulation around the region was not with the sole purpose of gathering, training, and inspecting a new police force. He was also being employed to produce evidence of King Lewanika's sphere of influence. As he circulated in the region, he compared the Lozi king's claims about the geographical extent of his rule against the accounts offered by local chiefs. The intention of Harding's superiors in London is clear. Their goal was to forge in the field an agent whose credibility was high, and whose local knowledge was so extensive as to be incontrovertible.

That Harding's superiors considered him successful in both diplomacy and border-scouting is evidenced by his later employment. He earned the trust of the king of the Lozi. In 1902, when the Lozi king attended the coronation of Edward VII, Harding was appointed as the Lozi escort.[29] When the western boundary of North-Western Rhodesia (nominally the western limits of the sphere of influence of the Lozi kingdom) was being arbitrated by an international boundary commission in 1903, Harding was called as a witness.[30] These tasks demonstrate the value that Harding's situated knowledge and locally-forged political relationships could have in meeting metropolitan goals.

However, Harding also used his influence to other ends. He drew on his situatedness and the perceived authenticity of his expertise to justify his actions in situations where he was *at odds* with his superiors. In August 1900, Harding and his patrol were dispatched to the north bank of the lower Zambezi valley. Africans had been migrating over the river to avoid paying new taxes being levied in the administrative districts that lay south of the river.[31] Harding's patrol was instructed to chase them back. However, Harding's local informers persuaded him that movement from one side of the river to the other was a historic pattern of mobility for the riverine communities. His sense of the traditions that preceded the establishment of colonial authority meant he was reluctant to enforce a strict tax boundary based on new colonial administrative geographies. His relationships with local informers generated a sense of justice that operated on a different logic than that of the newly established state. As a result, Harding pursued his instructions somewhat half-heartedly.

A second source of contention was the persistence of the relationships on which Harding's tubulous authority was built. The Lozi king saw Harding as an ally in his struggle against colonial machinations to strip him of his authority.[32] Harding regularly reminded the British authorities of the commitments that they had made to the Lozi. Harding re-iterated both the BSA's and the Colonial Office's statements about the Lozi king's sovereignty over taxation policy and the geographical extent of the Protectorate and his words carried weight.[33]

By April 1905, Harding's subordination had become more serious. His patrol was enrolled to assist with the collection of a new hut tax from African subjects. He was instructed to join local district officers in punishing those who resisted making payment. Again, Harding's understanding of African social practices prior to the establishment of colonial administration provoked a different perspective of what was practical or just than was held by the BSA directors. When Harding was asked to burn the huts of non-payers he refused. Writing directly to the High Commissioner in South Africa, Harding complained that it was useless to tax people who did not have money to give over. Harding's superior in Northern Rhodesia, Robert Coryndon, was intensely frustrated at being bypassed, and in a long letter bemoaned Harding's actions: "You have a great name for knowledge of the country and natives, and an adverse criticism from you would have a very bad effect on the Colonial Office estimate of my native administration."[34] That annoyance was compounded when Harding's position was upheld by the Colonial Office, against that of the BSA. This, finally, seems to have made Harding more of a liability than an asset to the BSA, and in 1906 he was effectively maneuvered into resignation.[35]

Clearly, Harding's personal political geography made him useful colonial instrument. His constant journeying across the region gave an air of authenticity to the knowledge that he passed upwards to the metropolitan center. This authenticity made his knowledge more powerful in the various colonial centers. Simultaneously, his mobility and networks enhanced the projection of his personal presence across the region and increased his status with the Lozi king and other African authorities. Yet, the political influence that Harding gained in this way could be, and was, deployed against his superiors.

Gielgud's Mobility

Valdemar Gielgud (?–1916), was another colonial officer whose personal authority sometimes overstepped the geographies and political ambitions of either the BSA or the Colonial Office. Gielgud first joined the BSA in 1890, before being dismissed as unfit in 1891. However, after fighting for the BSA in several conflicts, he was reinstated as a member of the company administration. From the late 1890s, Gielgud held posts in the BSA chartered territory south of the Zambezi, before being sent north of the river in 1900 to the Hook of Kafue. Although white traders and prospectors were well established, Gielgud was the vanguard of British authority and was sent with instructions to "report on the country and the conditions of the natives and to pave the way for peaceful occupation and administration."[36]

In some senses, Gielgud's approach to colonial administration was very different from Harding's. Gielgud was far quicker to take up violence, preferring (in his own words), a "show of force" over "logical argument and assurances."[37] Yet although Gielgud did not respect the existing laws and traditions of the peoples he encountered, he still relied on African socio-political geographies that pre-dated his arrival. Gielgud's reports and correspondence demonstrate that he was also, almost continuously, on the move. Like Harding, Gielgud's "patrols" were a series of visits to local political authorities: chiefs and headmen. However, rather than opening dialogue at each of these meetings, Gielgud simply pronounced his own authority and the new laws that took force under the colonial occupiers.

An even less likely scholar than Harding, Gielgud nonetheless produced maps that provided cartographic accounts of his activities and were returned with his reports. Although it has been suggested that Gielgud's patrol maps can be understood as early settlement maps of the Hook of Kafue, I would argue that this should not be the primary interpretation.[38] Gielgud's key goal was not defining or locating the occupants of the area. I would argue that his maps were spatial diagrams of political *pressure points*. Why so? Gielgud used the chiefs as pressure points to achieve control over the broader population of the Hook of Kafue. As well as promulgating British rule, he used his official influence over those chiefs in order to send Africans to work in Southern Rhodesia as migrant labourers.[39] He also deliberately turned the local social geographies against themselves, by always recruiting non-local Africans to his police escort, so that his unit would be less likely to show compassion or loyalty to the resident villagers.[40]

Although Gielgud's superiors were wary of his approach, he became a key element of negotiations over the new administrative geography of the area, and in those negotiations the strength of his position becomes clear. From 1901, there was discussion and disagreement over which administrative division the Hook of Kafue should join, the chartered colony North-Eastern Rhodesia, administered from Fort Jameson in the north, or North-Western Rhodesia, the colony built on the Lozi kingdom that was administered from Lealui in the south-west (see Figure 6.1).[41] Gielgud entered enthusiastically into the debate. Like Harding, Gielgud knew that his long presence in various sites across the Zambezi region made his knowledge reliable and indispensable.

> In writing the following [...] I am well aware that I am presuming to offer suggestions and express views on matters that do not directly fall within the scope of my official work here [...] in the belief that I am in a better position to judge the attitude of the natives in the Hook of Kafue than anyone else.[42]

Even more strikingly, his personality became a factor in these higher-level negotiations about the political future of the region. The BSA were attempting to draw up administrative boundaries that were a good "fit." Partly, this decision was predicated on whether the boundaries would be a good ethnic "fit" (i.e. an authentic grouping of the different "tribes" in the Hook of the Kafue region). However, discussion also turned on whether Gielgud's methods were a good "fit" for the differing political attitudes of the Administrators of North-Eastern and North-Western Rhodesia. Gielgud had become synonymous with the terrain in the Hook of Kafue. Bio-geography, narrative, and territory now coincided.

Eventually, however, like Harding, Gielgud became a thorn in the side of the higher ranks of colonial administration. Gielgud's militancy (and what was later characterized as paranoia about the threat of African violence) became a threat to the goals of the BSA and Colonial Office in the region. He was removed from his post. Gielgud maneuvered his dismissal more effectively than Harding and secured himself a position in which he could continue to press the same socio-political networks in the service of different masters. Sometime in 1903, Gielgud took as post as labor agent just south of the Zambezi, still deploying a coercive influence over political geographies, but now focused on using that influence to channel labor towards mines in the south.[43] By 1908, he was reinstated as a governmental administrator in Southern Rhodesia, working along a key route for labor travelling towards mining centers.[44]

Tying Bio-Geographies to Territory, Then Untying Them

In trying to understand what Gielgud's and Harding's maps were doing, we might want to explore the parallels between their journeying practices and those of European "explorers." There are several important commonalities. Firstly, Gielgud and Harding used routes and networks that pre-dated their arrival in the Zambezi region. They appropriated the infrastructure and resources of the local informants with whom they travelled, relying on their linguistic skills, mental maps, social relationships, even the physical pathways already carved into the landscape. The fourth item in the list of edicts that Gielgud pronounced to each chief that he met throws the dependence of these pioneer administrators into sharp relief: "You must treat white men well and not break agreements with them, nor lead them astray on wrong paths and so on."[45] Secondly, there is a parallel between Gielgud and Harding's epistemic practices and those of nineteenth-century explorers. The officers' reports emphasize the epistemic virtues of direct, situated experience in similar terms to those that explorers used to advocate their testimonies over the book-learning of armchair geographers.[46]

Gielgud's and Harding's work has a less comfortable parallel with exploration, however, where explorers' mapping is conflated with survey. As Edney explains (and more recently Naylor and Schaffer), survey operates through standardization. Through the flattening of local differences, survey offers a conceptual framework that stabilizes the definition of a territory.[47] Several scholars have argued that it is this standardization and stability offered by survey that has "allowed" the construction of modern states.[48] Yet it is important to pick out the dualities inherent to this process. Whilst surveyors might have been contributing to a project of standardizing and stabilizing territorial definitions, they were often simultaneously also pursuing their own agenda. Frequently explorers' accounts simultaneously constructed an abstracted geometrical conception of the terrain and a *narrative* for that terrain that celebrated their movement through it. In reporting their travels, such explorers were co-constituting their journeys with their biography, the maps traced both in space.[49]

Gielgud's and Harding's maps and reports also hold this duality. Read from one perspective, they are circulating items of geographical knowledge that allowed for a standardized representation of the Zambezi region in British colonial cartography. Yet in their reports, the agents were not rendering the field as "stable," but instead as somewhat "unstable." Those officers were using their maps and reports to emphasize the irreducible *particularity* of the areas under their jurisdiction, a particularity that could only be translated to the metropole through the ongoing co-location of their own person and the political space. The significance of administrative boundaries being organized around Gielgud's and Harding's activities shows the power of their brokerage and of their biographical narratives within territorial geographies.

From the perspective of the metropolitan center, the first years of the twentieth century represented a liminal moment. As a new colonial territory was being constructed, the political use of maps of the Zambezi basin was shifting from one paradigm to another. Under the earlier paradigm, the sketches that Gielgud and Harding made of their journeys were being read as evidence of political realities, a basis on which the British could make territorial claims. They were analyzed and recombined in documents such as the one in Figure 6.3, from 1901, which was used by the Colonial Office in preparation for the settlement of the Barotseland boundary with the Portuguese. The particularizing efforts of Gielgud and Harding and their bio-geographically-derived expertise were rhetorically useful to the authorities in London.

Figure 6.3 Map of the Barotse Kingdom (whole and detail). The figure shows different journeys in the Upper Zambezi region that had been reported by travelers. *(Continued)*

Source: National Archives U.K., MPG1/1048/3.

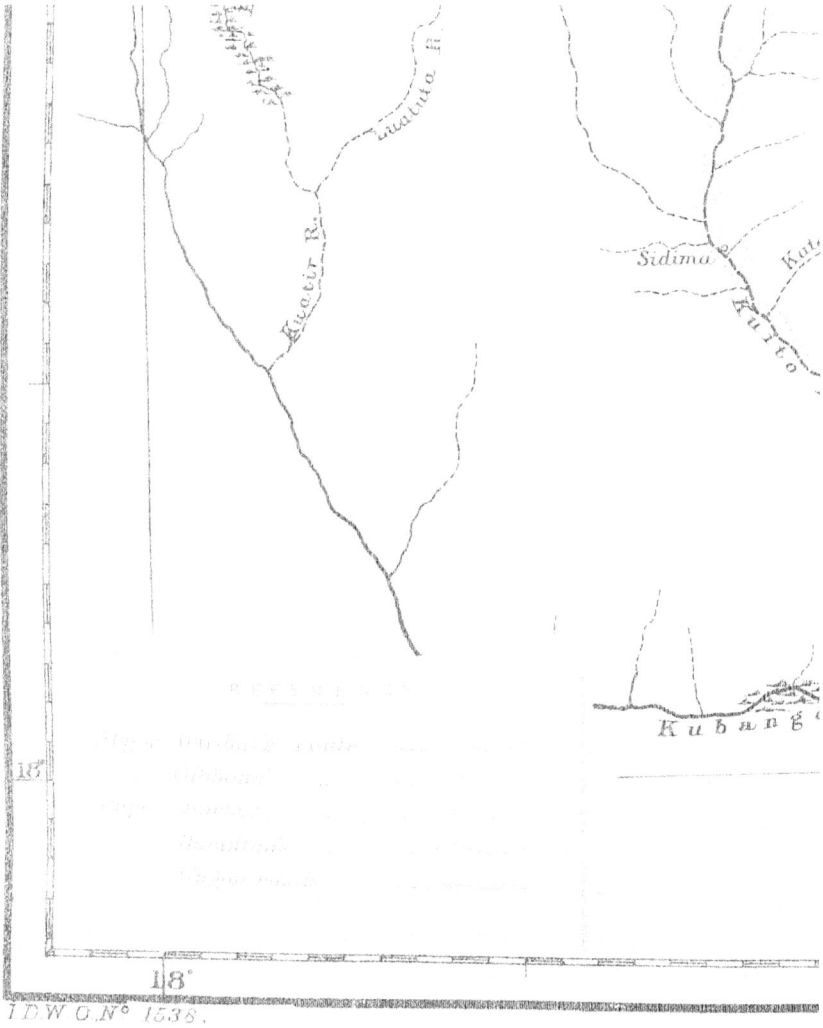

Figure 6.3 (Continued)

In the later paradigm, however, map documents were being used to *state*, rather than to claim: rather than "I saw," "he went," those later maps state "it is." In this paradigm of *stating,* depicting the territory as congruous with Gielgud's and Harding's bio-geographical narratives was dangerous, as it opened up the possibility of contestation. Admitting the variegated and historically contingent formation of the territory, which made state authority vulnerable, became a liability.[50] New maps of the colonial political territory eradicated the journeying that had produced the

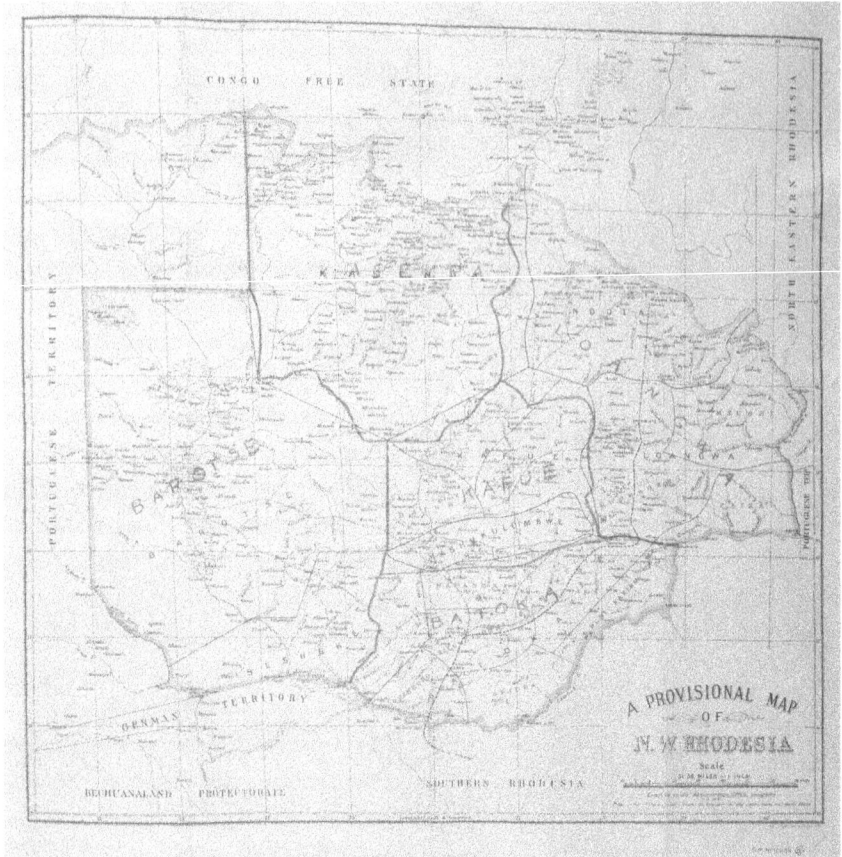

Figure 6.4 A Provisional Map of North-Western Rhodesia, 1908. This was the first published map of the colony that detailed it as a political territory.

Source: National Archives U.K, MR1/1830/5.

territory, as can be seen in Figure 6.4, the first map of the whole territory of North-
Western Rhodesia dating from 1908. Here routes that in Figure 6.3 are marked
with biographical information are now simply paths or roads, pre-colonial caravan
routes that inexpediently crossed over colonial territorial boundaries have disappeared, and there are no sites attached to individual's names, such as "Gielgud's"
camp at Mwenga River.

Bifurcating Genres: Occluded Mobility

It could be argued that what Gielgud's and Harding's bio-geographies reveal is
merely the usual course of events at the origin moment of a modern state territory, an account of the emergence of a bureaucratic apparatus and centralized government. Certainly, the first and second decades of the twentieth century saw an

increase in the layers of reporting and paperwork that mediated between the experience of governmental field agents and those in charge in London. The amount of direct contact between administrative officers posted to the Zambezi and top officials of the BSA and the Colonial Office decreased. Administrative posts in the new colony became more spatially fixed and were no longer so explicitly allied to individual persons.

Yet, this apparent standardization obscures power relationships that endured in the field. In the subsequent decades of the twentieth century, "patrols" became "tours." The almost-constant "touring" of administrative officers through the areas under their jurisdiction continued to be a vital part of colonial political practice in Northern Rhodesia, as across many British imperial territories.[51] Later generations of recruits to colonial administrative positions were no longer veterans of pioneer warfare, but were still required to be a "center of influence."[52] Throughout the colonial era, recruiters to colonial administration prioritized characteristics idealized as good-sense, leadership, and charisma over aptitude for meticulous record-keeping.[53] The structure of administration still favored the "discretion of the field over control from the centre," which was "protected by a claim to unique understanding of local conditions and effective control over information passed up official channels."[54] Mobility and governmental bio-geographies (grafted onto African social geographies) retained importance as colonial political technologies in Northern Rhodesia until Zambian independence in 1964.[55]

Whereas Gielgud and Harding were producing maps that could be read in two ways, as either building state territory or charting bio-geographies, subsequent colonial map production bifurcated into two separate genres. Local colonial administrative offices continued to produce maps, as each agent reported their tours with crude small-scale sketches of their journeys.[56] Biography and territory were still intertwined in the daily execution of colonial administrative practice. Crucially, however, within a decade those maps were no longer sent to London, or even the colonial capital in Livingstone. Maps that carried a pragmatic recognition of the *de facto* tubulous sovereignty of the colony only circulated as far as District or Provincial level. On the other hand, centrally-produced maps of Northern Rhodesia used European conventions to depict uniform areas of jurisdiction. They failed to reveal the ongoing connections between bio-geographies and political territory. They occluded rather than revealed the mechanisms through which the state was being constructed and maintained. The result has been the production of two distinct archives, one in London, and one in Lusaka, giving two very different accounts of colonial political geographies. I would suggest that this divergence of cartographic form and audience was vital in enabling the British to maintain the appearance of rational bureaucratic governance, whilst delivering the "decentralized despotism" of indirect rule.[57]

Conclusion

By taking the bio-geographies or movement histories of Gielgud and Harding seriously, we are offered new ways to think about the role of geographical knowledge in the generation of colonial territory. We can better understand the dissonance that was created in the British mindset as they simultaneously attempted to produce a centralized bureaucracy whilst controlling colonized peoples through indirect rule. We see the disjuncture between the forms of geographical knowledge that typified these two different impulses in governmental practice, and a similar disjuncture in the patterns of the circulation of that knowledge.

There was, apparently, a limit to what the colonial and metropolitan centers wanted to know about the socio-political geographies of Northern Rhodesia. Those geographies, at one point in time visible and fervently discussed in London, sank down through the sediments of new layers of bureaucracy. Nonetheless, political control over the territory continued to rely on administrators exerting a negotiated, relational influence over African networks and infrastructures.[58] The personal, mobile, and relational exercise of power and the corresponding biographical, linear, and narrative forms of documentation, which have previously been described as "frontier" forms of paperwork, can be seen to have sat right at the heart of colonial political modalities.[59] The cartographic evidence of those mobile political geographies, whilst available, remained deliberately local, tied to the persons of the "man on the spot."

Notes

1 Jeffrey Stone, "An Early Map of the Hook of Kafue," in R. C. Bridges, ed., *An African Miscellany for John Hargreaves* (Aberdeen: Aberdeen University African Studies, 1983), 93–96.
2 Valdemar Gielgud, Mwenga River, "Report (2)" (9 October 1901), D3287 BSA/8/44, Derbyshire Record Office.
3 Jeffrey Stone, "The District Map: An Episode in British Colonial Cartography in Africa, with Particular Reference to Northern Rhodesia," *The Cartographic Journal* 19, no. 2 (1982): 104–112; Stone, "An Early Map of the Hook of Kafue"; Jeffrey Stone, *A Short History of the Cartography of Africa* (New York, NY: E. Mellen Press, 1995).
4 Keith Breckenridge, *Biometric State: The Global Politics of Identification and Surveillance in South Africa, 1850 to the Present* (Cambridge: Cambridge University Press, 2014); Matthew Stuart Hull, *Government of Paper: The Materiality of Bureaucracy in Urban Pakistan* (Berkeley, CA: University of California Press, 2012); Deborah Poole and Veena Das, eds., *Anthropology in the Margins of the State* (Oxford: James Currey, 2004); Simon Schaffer et al., eds., *The Brokered World: Go-Betweens and Global Intelligence, 1770–1820* (Sagamore Beach: Science History Publications, 2009).
5 Bernard S. Cohn, *Colonialism and Its Forms of Knowledge: The British in India* (Princeton, NJ: Princeton University Press, 1996); Raymond B. Craib, *Cartographic Mexico: A History of State Fixations and Fugitive Landscapes* (Durham, NC: Duke University Press, 2004); Matthew H. Edney, *Mapping an Empire: The Geographical Construction of British India, 1765–1843* (Oxford: Oxford University Press, 1997); Timothy Mitchell, *Rule of Experts: Egypt, Techno-Politics, Modernity* (Berkeley: University of California Press, 2002).
6 Sara Berry, *Chiefs Know Their Boundaries: Essays on Property, Power, and the Past in Asante, 1896–1996* (Portsmouth: Heinemann, 2001).
7 Alistair Tough and Paul Lihoma, "The Development of Recordkeeping Systems in the British Empire and Commonwealth, 1870s–1960s," *Archives and Manuscripts* 40, no. 3 (2012): 191–216.
8 Bruce Berman, *Administration and Politics in Colonial Kenya* (New Haven, CT: Yale University, 1974), 27–28.

9   High Commissioner for South Africa, "Letter to the Secretary of State for the Colonies" (11 November 1907), CO 879/95/872, National Archives, U.K.

10 Bruce Berman, "Structure and Process in the Bureaucratic States of Colonial Africa," *Development and Change* 15, no. 2 (1984): 161–202.

11 Jeffrey Herbst, *States and Power in Africa: Comparative Lessons in Authority and Control* (Princeton, NJ: Princeton University Press, 2000), 82. For an interrogation of how that de-centredness could be harnessed by colonial scientists see Helen Tilley, *Africa as a Living Laboratory: Empire, Development, and the Problem of Scientific Knowledge, 1870–1950* (Chicago, IL: University of Chicago Press, 2011).

12 See for example: Herbst, *States and Power in Africa*; Jan-Bart Gewald, Sabine Luning, and Klaas van Walraven, "Motor Vehicles and People in Africa: An Introduction," in Jan-Bart Gewald, Sabine Luning, and Klaas van Walraven, eds., *The Speed of Change Motor Vehicles and People in Africa, 1890–2000* (Leiden: Brill, 2009), 1–20; Allen M. Howard, "Nodes, Networks, Landscapes and Regions: Reading the Social History of Tropical Africa 1700s–1920," in Allen M. Howard and Richard Matthew Shain, eds., *The Spatial Factor in African History: The Relationship of the Social, Material, and Perceptual* (Leiden: Brill, 2005); Clapperton Chakanetsa Mavhunga, *Transient Workspaces: Technologies of Everyday Innovation in Zimbabwe* (Cambridge: MIT Press, 2014); Malyn Newitt and Corrado Tornimbeni, "Transnational Networks and Internal Divisions in Central Mozambique," *Cahiers d'études Africaines*, no. 192 (2009): 707–740; Marion Wallace, "Personal Circuits: Official Tours and South Africa's Colony," *Journal of Southern African Studies* 41, no. 3 (2015): 635–652.

13 Although the history and legacy of colonial labour migration has seen much scrutiny, the colonial use of mobility as an administrative tool has not been similarly considered. Scarlett Cornelissen, "Migration as Reterritorialization: Migrant Movement, Sovereignty and Authority in Contemporary Southern Africa," in Patrick Chabal, Ulf Engel, and Leo de Haan, eds., *African Alternatives* (Leiden: Brill, 2007), 119–143; Darshan Vigneswaran and Joel Quirk, eds., *Mobility Makes States: Migration and Power in Africa* (Philadelphia: University of Pennsylvania Press, 2015).

14 Bérénice Guyot-Réchard, "Tour Diaries and Itinerant Governance in the Eastern Himalayas 1909–1962," *The Historical Journal* 60, no. 4 (2017): 1025.

15 Lauren Benton, "Spatial Histories of Empire," *Itinerario* 30, no. 3 (2006): 19–34.

16 Schaffer et al., *The Brokered World*.

17 Paul Carter, *The Road to Botany Bay: An Exploration of Landscape and History* (New York: Knopf, 1987); D. Graham Burnett, *Masters of All They Surveyed: Exploration, Geography, and a British El Dorado* (Chicago, IL: University of Chicago Press, 2001); Guyot-Réchard, "Tour Diaries and Itinerant Governance in the Eastern Himalayas 1909–1962"; David Ludden, "The Process of Empire: Frontiers and Borderlands," in Peter Fibiger Bang and C. A. Bayly, eds., *Tributary Empires in Global History* (London: Palgrave Macmillan, 2011), 132–150; Philip E. Steinberg, "Sovereignty, Territory, and the Mapping of Mobility: A View from the Outside," *Annals of the Association of American Geographers* 99, no. 3 (2009): 467–495.

18 Paul Carter, *Dark Writing: Geography, Performance, Design* (Honolulu: University of Hawaii Press, 2009), 20.

19 Ibid., 28.

20 The term bio-geography here held refers to the "movement history" of an individual rather than biogeography as the distribution of biological entities in space and time.

21 Carter, *The Road to Botany Bay*, 71.

22 Ann Laura Stoler, *Along the Archival Grain: Epistemic Anxieties and Colonial Common Sense* (Princeton, NJ: Princeton University Press, 2010).

23 See Michel de Certeau, *The Practice of Everyday Life*, trans. Steven F. Rendall (Berkeley: University of California Press, 1984), 97. Thus allowing the British imperial state to forget the role of what Scott describes as "metis" in colonial governance.

James C. Scott, *Seeing Like State: How Certain Schemes to Improve the Human Condition Have Failed* (New Haven, CT: Yale University Press, 1998).

24 Tim Wright, "1752. Colin Harding of the 2nd Regiment, King Edward's Horse," *Journal of the Society for Army Historical Research* 83, no. 335 (2005): 255–257.

25 Tim Wright, *The History of the Northern Rhodesia Police* (Bristol: British Empire and Commonwealth Museum, 2001), 39.

26 Colin Harding, *In Remotest Barotseland: Being an Account of a Journey of over 8,000 Miles through the Wildest and Remotest Parts of Lewanika's Empire* (London: Hurst and Blackett, 1904); Colin Harding, *Far Bugles* (London: Simpkin Marshal, 1933); Colin Harding, *Frontier Patrols. A History of the British South Africa Police and Other Rhodesian Forces* (London: C. Bell & Sons, 1937).

27 Diary of William Hallett Harding MS. 5433, Bodleian Library, Oxford, U.K.

28 Wright, *The History of the Northern Rhodesia Police*, 42.

29 Harding, *Far Bugles*.

30 Parliament of Great Britain, *Award of His Majesty the King of Italy Respecting the Western Boundary of the Barotse Kingdom* (London: Printed for H.M.S.O. by Harrison and Sons, 1905).

31 JoAnn McGregor, *Crossing the Zambezi: The Politics of Landscape on a Central African Frontier* (Woodbridge: James Currey, 2009); Kenneth Powers Vickery, *Black and White in Southern Zambia: The Tonga Plateau Economy and British Imperialism, 1890–1939* (New York, NY: Greenwood Publishing Group, 1986); Wright, *The History of the Northern Rhodesia Police*, 42.

32 "Letter from Robert Coryndon, Kalomo, to the High Commissioner for South Africa" (20 October 1904), CO417/480, National Archives, U.K.

33 See for example: "Letter from Colin Harding, to Robert Coryndon, Sesheke" (7 December 1902), CO417/480, National Archives, U.K.

34 "Letter from Robert Coryndon, Lealui to Colin Harding, Kalomo" (5 June 1905), CO417/480, National Archives, U.K.

35 "Statement of Facts Relative to the Resignation of Colonel Colin Harding CMG. Late Commandant of the Barotse Police; Also Late Acting Administrator and British Resident, Barotseland, North-West (Sic) Rhodesia (Anonymous)" (1909), CO417/480, National Archives, U.K.

36 S. R. Denny, "Val Gielgud and the Slave Traders," *Northern Rhodesia Journal* III, no. 4 (1957): 334.

37 "Letter to Cecil Rhodes, from Val Gielgud, Hook of Kafue" (10 February 1901), MSS Afr. S. 228 (1), Bodleian Library, Oxford.

38 Gielgud's maps are characterized in this way by Jeffrey Stone Stone, "An Early Map of the Hook of Kafue"; Stone, "The District Map"; Jeffrey Stone, "Maps as Sources in Demographic Studies of Africa's Past, with Particular Reference to Zambia," in Bruce Fetter, ed., *Demography from Scanty Evidence: Central Africa in the Colonial Era* (Boulder, CO: Lynne Rienner Publishers, 1990), 161–182.

39 "Letter to Cecil Rhodes, from Val Gielgud, Hook of Kafue."

40 Val Gielgud, Mwenga River, "Private Report on Northern Rhodesia" (14 May 1902), D3287 BSA/8/60, Derbyshire Record Office.

41 "Memorandum for the Directors of the BSA from Robert Codrington, Administrator of North-Eastern Rhodesia" (31 December 1901), D3287/BSA/8/49, Derbyshire Record Office.

42 Valdemar Gielgud, Mwenga River, "Report (1)" (26 August 1901), D3287 BSA/8/44, Derbyshire Record Office.

43 Lewis H. Gann, *A History of Southern Rhodesia: Early Days to 1934* (London: Windo and Chattus, 1965).

44 Mavhunga, *Transcient Workspaces*, 105.

45 Valdemar Gielgud, Mwenga River, "Report (2)." Apparently, Gielgud was "led astray" within the first weeks of his arriving in the Hook of Kafue when he attempted to make

his way over to another colonial outpost at Chipepo. Denny, "Val Gielgud and the Slave Traders," 331.

46 Charles W. J. Withers, "Mapping the Niger, 1798–1832: Trust, Testimony and "Ocular Demonstration" in the Late Enlightenment," *Imago Mundi* 56, no. 2 (2004): 170–193.

47 Edney, *Mapping an Empire*; Simon Naylor and Simon Schaffer, "Nineteenth-Century Survey Sciences: Enterprises, Expeditions and Exhibitions," *Notes and Records: The Royal Society Journal of the History of Science* 73, no. 2 (2019): 135–147.

48 See for example: Stuart Elden, *The Birth of Territory* (Chicago, IL: University of Chicago Press, 2013); Mitchell, *Rule of Experts*; John Kenneth Noyes, *Colonial Space: Spatiality in the Discourse of German South West Africa 1884–1915* (Chur: Harwood Academic Press, 1992).

49 Carter, *The Road to Botany Bay*; Burnett, *Masters of All They Surveyed*.

50 Berry, *Chiefs Know Their Boundaries*.

51 Jan-Bart Gewald, "People, Mines and Cars: Towards a Revision of Zambian History 1890–1930," in Jan-Bart Gewald, Sabine Luning, and Klaas van Walraven, eds., *The Speed of Change Motor Vehicles and People in Africa, 1890–2000* (Leiden: Brill, 2009), 21–47; Anthony Kirk-Greene, *Symbol of Authority: The British District Officer in Africa* (London: I. B. Tauris, 2006).

52 High Commissioner for South Africa, "Letter to the Secretary of State for the Colonies" (11 November 1907), CO 879/95/872, National Archives, U.K.

53 Chris Jeppesen, "'Sanders of The River, Still the Best Job for a British Boy': Recruitment to the Colonial Administrative Service at the End of Empire," *The Historical Journal* 59, no. 2 (2016): 469–508.

54 Berman, "Structure and Process in the Bureaucratic States of Colonial Africa," 176 and 181.

55 Elizabeth Haines, *A Colonial Cartographic Economy: The Contested Value of Mapping in Northern Rhodesia, 1915–1955* (London: University of London, 2016).

56 Stone, "The District Map."

57 Mahmood Mamdani, *Citizen and Subject: Decentralized Despotism and the Legacy of Late Colonialism* (Oxford: Oxford University Press, 1997).

58 Berman, "Structure and Process in the Bureaucratic States of Colonial Africa."

59 Guyot-Réchard, "Tour Diaries and Itinerant Governance in the Eastern Himalayas 1909–1962."

7 The Global Dimensions of the Rome Zoological Garden and Italian Colonialism in Africa

Mauro Capocci and Daniele Cozzoli

Introduction: Urban Spaces, Zoological Gardens, and Colonialism

Throughout the long nineteenth century, the birth of the modern idea of the nation and the consequent creation of the nation-state, as well as the great expansion of the circulation of goods, people, ideas, behaviors and cultural products, eased by the technological innovations of the telegraph, the railway, the steamship and the electrification of cities, contributed to make the world more homogenous in a number of respects.[1] As stressed by Christopher Bayly, this process did not consist of a simple exportation of Western patterns to the rest of the world. In some cases, different agencies contributed to shape material and immaterial processes. Both Hamburg and New York capitalists and old-style Chinese entrepreneurs, for instance, contributed to the increase of world trade in China seas and West Asia.[2] In other cases, patterns spread quickly from one region to the rest of the world. Chinese pottery flooded Europe since the eighteenth century[3], while the European Opera (originated in Italy in the seventeenth century) became the universal language of music and led to the creation in the nineteenth century of opera theatres even in countries, such as China and Japan, which had a strong indigenous opera tradition.[4] Such changes deeply affected societies worldwide and had consequences in the setting of urban spaces. The renewed European efforts in colonial expansion after the "Scramble for Africa" also created or redesigned some urban spaces, such as colonial hospitals, world fairs, and zoological gardens. In the nineteenth century, animals ceased to be shown in Royal menageries and were incorporated into public spaces, where a larger public could watch them.[5] Historians usually point at the creation of the menagerie within the Jardin des Plantes under the auspices of the Muséum National d'Histoire Naturelle in Paris (1793) as the starting point of a new idea of living animal collections, to be used for science and for public entertainment, and not as the solace to a royal court nor as a travelling circus.[6] Naturalist Bernardin de Saint-Pierre, endorsing the creation of the menagerie in Paris, stated that it was necessary for the "propriety and the dignity of the nation" and for the "political relations": the new Republic needed a place for keeping the living gifts from other countries.[7] On a different tune, science was the prime motive for the opening of the London Zoo in 1828, created by the London Zoological Society and open to the public after two decades: it was the venue for the first encounter between a non-human

DOI: 10.4324/9780367814540-8

ape – the orang-outan Jenny – and Charles Darwin in 1838.[8] Other meanings are attached to the development of zoos in the nineteenth century. The Amsterdam Zoological Society and its associated zoological garden were created in 1838 and have been considered the result of the "new bourgeoisie using their financial capital to create cultural institutions":[9] a public display of their collective values, including colonial pride. The Berlin Zoo opened its gates in 1844, following yet another trajectory. It was the result of the naturalists and explorers Alexander von Humboldt and Martin Lichtenstein pressing the Kaiser Friedrich Wilhelm IV, arguing for the prestige of the nation and the several benefits that zoo would provide to German science and agriculture. However, German industrialists and rich bourgeoisie did not contribute, and the initial offer of shares to support the opening of the zoo failed miserably.[10] As a result, for over two decades, the Berlin Zoo was mostly supported by the Kaiser himself. Throughout the nineteenth century, zoological gardens were created all over the world: Melbourne (1857), Moscow (1864), Tokyo (1882), El Cairo (1891), Buenos Aires (1888), Rio de Janeiro (1888), New York (1899), Pretoria (1899). A bourgeois public felt the urge of watching exotic animals they have previously only seen in drawings, but also of interacting with them in some ways. The spreading of Darwinism stimulated the interest and the curiosity for apes: the public wanted to see what were perceived as their "ancestors."[11]

Hagenbeck in Rome

At the beginning of the twentieth century, Carl Hagenbeck (1844–1913) inspired a new model for the zoological gardens, mirroring a new relationship between the public and the exotic.[12] Hagenbeck, a son of an animal tradesman, had become the most important animal dealer in the world, with a global network of trappers and hunters collecting live specimens for travelling circuses and zoological gardens alike. In Hagenbeck's park, created in Stellingen (at the time still outside Hamburg) in 1907, animals were not arranged according to any systematic method, but were exhibited without cages in large "panoramas" recreating some exotic habitat. Individuals from different species were often grouped together and their spaces were separated from the public by moats instead of cages. The public participated in an illusion of "natural habitat." The innovative arrangement created a considerable stir, attracting a large public to Hagenbeck's Tierpark.

It is not thus entirely surprising that Hagenbeck was hired to plan the new Rome Zoological Garden. Indeed, in 1907 the high-rank civil servant Riccardo Anellis proposed to the City Council the creation of a zoological garden in Rome. Anellis was the leader of a group of entrepreneurs supported by the Banco di Roma (a Vatican-based bank that bought 13% of the shares). Hagenbeck provided most of the animals and invested heavily in the company shares, being the second largest shareholder.

The location chosen for the zoo was the Villa Borghese, a large estate north of the historical city. Rome was growing fast in the second half of the nineteenth century, like many other cities (especially port cities) around the world. When Rome became the Capital of the Kingdom of Italy in 1871, it was still largely comprised within the ancient Roman circle of walls, interspersed with several undeveloped areas often used for agriculture. The development of these areas came about almost abruptly in the last three decades of the century, with the reduction or the complete annihilation of several green expanses to make room for new buildings. The fate of Villa Borghese partly reflects the urban history. Owned by the powerful Borghese family, in the later decades of the nineteenth century, the park was already open to the public for leisure and amusements. Before the opening of the Giardino Zoo-logico, it even hosted – in two different ages – a menagerie. The first was created in the seventeenth century by Scipione Borghese. The second one was a small exhibition of exotic animals on the public grounds of the park, open from 1893 to 1911 under the name "Acclimatization Garden," echoing the privately owned French amusement park in the Bois de Boulogne.[13] The Villa – and the wonderful art collection it hosted – was the object of a long controversy between the City and the Borghese about the right to use the area as a common, sparing it from the furious development that was swallowing neighboring districts. Eventually, the park was acquired by the State in 1901, and immediately donated to the City of Rome (except for the art gallery). Adjacent to the town's historical core, the Villa conveniently bordered with the countryside, while being close to other developing areas. Already an established place for public entertainment, with large expanses to be modelled according to necessities, it was a perfect spot for the creation of the Zoological Garden.

The City Council accepted Villanis' proposal, based on the operation of a joint stock company founded in 1909: the *Società Anonima per l'impianto e l'esercizio del Giardino Zoologico in Roma* (Anonymous Society for the establishment and the operation of the Zoological Garden in Rome). The company obtained a 45-year lease on the land, a 11 hectares of agricultural estate on the northern border of the park. The first president of the company was the politician and economist Giorgio Sonnino (1844–1921), born in Alexandria, to a Jewish merchant and an English woman who raised their children as Anglicans. His younger brother Sidney Sonnino served as Treasury minister (1893–1896), Prime minister and Internal affairs minister (in two short spells, in 1906 and 1909), and Foreign affairs minister (1914–1919). Giorgio Sonnino was among the founding members of the Italian Geographical Society (1867) and as a member of the Parliament, he was appointed in the committee for the Italian Somaliland legislation in December 1906. The political prominence of the two brothers was most likely central to convince several senators to buy shares of the company and to secure the support for the initiative.

The promoters of the Zoo mostly appealed to the educational benefit of such an institution. However, according to the director of the Garden Division of the City Council, the Zoo would have been also useful to fight the "absolute ignorance we have about our Italian dominions in Africa."[14] In his opinion, a part of the Zoo should have been devoted to zoological and botanical specimens from the Eritrea colony.

Hagenbeck's plan included exotic buildings, though not necessarily consistent with the animals' origin. This approach to zoo architecture had been already implemented in Berlin Zoo since the 1870s, with the antelopes and the elephants living not in cages but in magnificent Oriental-style palaces, as if they were humans. This style was preferred because – in the words of the director – "everything Moorish has been seen as ultramodern"[15] ever since the king of Wurttemberg's Moorish castle in Stuttgart and the grandiose Suez Canal in Egypt. In Hagenbeck's opinion, the zoo was a window in the worlds of the "others," including humans. Nonetheless, the Berlin Zoo with its Elefantenhaus and Antilopenhaus transmitted to the public a positive and somehow reassuring image of the East. Yet, it imposed a simplistic view of the East that ultimately vehiculated the idea of backwardness of the Eastern civilizations that savage animals associated with Eastern buildings implicitly transmitted to the Berlin public.[16] In Rome, within the limited space of the park, these worlds could be recreated – with some scientific respectability – for the entertainment and the education of the public, with the hope to replicate the huge success of Berlin Elefantenhaus and Antilopenhaus. However, the exotic architecture was limited by budget restrictions. Still, it is worth underlining that the "pachyderm house" built in Rome had an "Egyptian design," though it hosted elephants from Ceylon, India, and German East Africa. Furthermore, the housing also accommodated an African rhino, a hippo, and an American tapir. The Egyptian design made no scientific sense, just like the term "pachyderm house": the only shared trait among those animals was the vegetable-based diet. A guide of the zoological garden explained that, although the term "pachyderm" was not scientifically accurate, animals in the pachyderm house were grouped together for practical reasons.[17] In the nineteenth-century zoos around the world, the relationship with humans was central for certain kinds of animals, and elephants were among those.[18] In Rome, the elephant Toto (originally M'toto, "child" in Swahili) became an attraction. Imported from German East Africa by Hagenbeck, Toto used to walk freely in Dar-es-Salaam, "where he has brought with calm dignity advertisement posters."[19] In order to increase the attractiveness of the park, a number of shareholders of the zoo even suggested that visitors could ride Toto and his fellow elephants.[20] Unfortunately, in 1921 Toto killed the zoo veterinarian and in 1928 murdered his guardian, proving how illusory was the image of a special elephant/human relation.[21] Four years later he had become in the press "the ferocious African elephant."[22] The "Hagenbeck system" typically also included humans as exotic species,[23] but no such display took place in Rome.[24] Nevertheless, the idea was periodically taken into consideration.

The colonial theme for the zoo did not become a major focus until Benito Mussolini took the power. The Italian dominions in Africa were mostly unknown to the public, and notwithstanding the efforts of several institutions, such as the Italian Colonial Institute (ICI), colonies were not attracting significant Italian emigration, nor were they contributing to the Italian economy. In the words of the above-quoted Severi "colonies are subject of conversation only when disasters and regrettable accidents occur, and this makes our dominions rather suspicious."[25] Furthermore, it was Hagenbeck that provided the animals, while the zoo – nor anybody else – was not willing to sponsor costly expeditions to collect live animals. Only a very small number of animals came from the colonies: two lions were donated, aptly named Eritrea and Negus. The first one was donated by the governor of the eponymous Italian colony, Giuseppe Salvago Raggi, and the second one by the wife of

Marquis Centurione Scotto (a member of the board of the Zoo company, and owner of 275 shares).[26] It is also noticeable the absence of any cooperation with the Italian Geographical Society and the ICI, notwithstanding the fact that Sonnino was a member of both institutions, and that several MPs were involved with the ICI.[27]

The venture soon started to lose momentum. The grand Hagenbeck project was expensive, and all the infrastructures had to be built from scratch, including electricity and water supplies, with no help from the city or the government. Even without ostentatious buildings (compared to other zoos), the viability of the project was dubious, and already in 1910, before the opening the shareholders voted for a change. Villanis and Sonnino were removed from the Board of Trustees, and the amateur ornithologist, pioneer photographer and motorist, Francesco Chigi, from an ancient and noble Roman family, was appointed as president. According to some shareholders, the trigger for the revolution was the project of a Luna Park within the Zoological Garden. As detailed by Francescangeli and Gippoliti, the proposal was made by an American entrepreneur but encountered Hagenbeck's opposition, as well as a few bureaucratic obstacles.[28]

The Zoological Garden finally opened – a few days late on schedule – on 5 January 1911, in time to be part of the celebrations for the 50th anniversary of the proclamation of the Kingdom of Italy.[29] The Zoo was Hagenbeck's creature, focussed on animals and their display within a "natural" environment, and directed by Theodorus Knottnerus-Meyer, zoologist and Hagenbeck's collaborator in Stellingen. Everybody was hoping that the zoo would benefit from the mass of tourists attracted to the city, but expectations were ill-placed. Since its first year the venture was in trouble.[30] In 1913, an anonymous pamphlet published by a group of shareholders of the zoo enlisted a number of motives of the failure. No public transport was provided for the zoo until 1913, and the managers were – according to some shareholders – unable to attract the public, with poor advertising and bad relationship with members of the press. Furthermore, when the zoo opened, the infrastructure was far from being complete, resulting in a bad experience for the visitors.[31] The animals suffered too: the wonderful "Noah's Ark" envisioned by Hagenbeck proved lethal for several of them, with many succumbing to diseases and poor management.[32] The zoo never fulfilled Hagenbeck's desires, since the arena intended for animal and human shows – a beloved attraction in other zoos and a large source of revenue for Hagenbeck's firm – was not created. The restaurant quickly declined, too, partly because the director Knottnerus-Meyer never allowed the restaurant to work beyond zoo's opening hours, even asking the restaurant patrons and suppliers to pay for the entrance ticket. A financial disaster waiting to happen, the company entered a debt arrangement procedure in 1912.[33] The restaurant was already closed, and its grand building was let to a movie production company. Looking at the failure, it is interesting to note that the fascination for all things exotic and colonial was probably not very strong neither in the Roman public nor in Italian popular culture. One of the reasons is probably the relative youth of the nation and its wide cultural diversity. In fact, one of the key features of 1911 national jubilee in Rome was the Ethnographic Exhibition, devoted not to faraway peoples, not even to the Italian colonies in East Africa, but to internal regional cultures, whose diversity was still striking – and a pride? – to Italians themselves. The colonial themes were instead displayed in the World Fair in Turin, held at the same time as the Ethnographic Exhibition in Rome, where African villages from Eritrea and Somalia were a great attraction. Quite tellingly, several observers used the same words to describe the colonial and the Italian "barbarians."[34] The capital of the young nation was probably not suitable for the exhibition of alien people when the newly created "Italians" were still foreign to each other.

In 1913, the company reached the budget balance. Nonetheless, its crisis was irreversible, and in the following year, it filed for bankruptcy, entering a long period of neglect.[35] In 1917 the City Council took the estate back, buying animals, buildings and anything of interest. However, during and after WWI the zoo was not among the top priorities.

It is also worth noting that neither the scientific nor the zoo-technical aspects were central. Criticisms were directed at Hagenbeck's zoo concept. In Italy as well as elsewhere, a number of zoologists and naturalists did not appreciate the absence of systematic order in the animal display and the emphasis on the shows.[36] And in Rome, universities and academics were not really involved in the establishment of the zoo. Although Francesco Chigi (since 1910 the president of the zoo company) was a member of the Rome-based Italian Zoological Society,[37] the latter association only paid attention to the zoo as a steady source of dead animals, to be added to Rome University zoological collections.

The creation of the zoo in Rome meant the participation in a horizontal network that included modern cities – the one that was connected by Hagenbeck's commercial relationships and by his idea of human/animal exhibitions.[38] On the other hand, it was also struggling to become the summit of a vertical hierarchy. For several reasons, the public in the capital of the Italian nation proved to be "reluctant" to the cultural entertainment that worked elsewhere in the country and abroad. At the same time, the financial elite pushing for the zoo did not catch up with the imperialist wind that was sweeping Italy and its politics.[39]

The Creation of the Empire, Public Sphere, and Public Spaces in Fascist Italy

The advent of Fascism in 1922 opened up a new page in the history of the zoo. As stressed by Sabino Cassese, the Fascist regime did not radically change the liberal state. By slightly modifying the existing legislation, Fascism created a much more authoritarian state, which suppressed political pluralism. Pre-Fascist liberal Italy was an authoritarian state tempered by some liberal institutions. On the other hand, a number of innovations introduced by Fascism, such as the organization of the work of the parliamentary assembly in smaller committees, was maintained by the post-WWII Republic.[40] Historians still debate if Mussolini's regime can be classified under the category of totalitarianism. Emilio Gentile stresses that Fascism aimed at creating a "new man"; therefore, it intervened in the private sphere of citizen life. Fascism, for instance, abrogated the handshaking and substituted it with the Roman salute, or prohibited the use of foreign words and replaced them with *ad hoc* Italianized terms. Anti-Fascists often made fun of these measures as well as of the pompous rhetoric of the regime. Those changes had, however, a clear-cut aim: an in-depth anthropological transformation of the character of the Italian people. Fascism created a "political religion" with its cult, its rites and language, and also with its "cathedrals": the public spaces where the political religion unfolded.[41] At the same time, although deprived of power, previous institutions, such as the Monarchy and the Senate were maintained. The Fascist militia was absorbed by the State. But the State and the Party never identified. Moreover, in 1929 the signing of the Lateran Pacts made Catholicism the official religion of the State and protected Catholic associations and non-political organizations. Fascism was not in this sense a totalitarian regime. As a consequence of this complex picture we just outlined, Fascism was constantly seeking consensus, therefore allowing to some extent the formation – if not of what Habermas defined a public sphere – at least of a space in which different and opposed views coexisted and a limited debate was possible. It is worth noting that this fact was not in contrast with the totalitarian aspect of Fascism: the will of creating a "new man."

It was within this complex framework that the zoo gained momentum. Since the early years of his mandate as premier, Mussolini looked at the zoo as an instrument of propaganda and connected it to the colonial venture. At the same time, the zoo could directly address the public opinion, whose approval was needed, and also for the Fascist totalitarian project. The human/animal relation, therefore, played a part in Mussolini's attitude. On 4 March 1924, the *Corriere della Sera* reported that the day before journalists were invited to visit the zoo and that they were shown its renovation project. The same newspaper also reported that the Ministry of the Colonies was committed to provide animals and that in the evening Mussolini showed up and paid a visit to the lioness he donated to the zoo.[42] The 1924 was the crucial year of the rise of Fascism. It is in a way astonishing that one month before the general elections – while violence of the Fascist paramilitary groups flared up – Mussolini paid so much attention to restructuring the zoo and to its animals. The year before, in 1923, Mussolini had donated to the zoo the lioness Italia, a gift from a circus. Mussolini used to visit her and portraits of *il Duce* with the lioness were published in newspapers. When Italia gave birth to 3 puppies, the *Giornale d'Italia* invited its readers to make proposals for their names. The puppies were named Bebè, Ninì and Totò, from the syllables of Mussolini's first name Be-Ni-To.[43]

The new Rome Governor engaged in a new effort in using the zoo as a vehicle for the Imperial propaganda of the Fascist regime. With the new management, the zoological garden eventually connected with the Ministry of the Colonies in order to have a steady flow of animals from Africa. The zoo, indeed, obtained preemptive rights over any animal captured under any circumstance in Italian East Africa.[44] In the meantime, in 1925, the Azienda del Giardino Zoologico was created and in 1930 the Count Guido Suardi was appointed commissioner of the Azienda. The creation of an aquarium was also considered since 1928.[45]

At the beginning of the 1930s, Mussolini envisaged renovating Rome Zoological Garden. On 29 November 1931, Suardi sent to il Duce a detailed project. Under the new perspective, the zoo was to be turned into an eminently cultural and scientific institution: therefore, the focus shifted from the number of specimens to the number of species. Indeed, 300 new species were needed, 80 of which were difficult to obtain. Between 1925 and 1930 species had already grown up from 345 to 412.[46] The number of visitors also increased.[47] Although the report stressed that animals that could attract visitors had to be chosen, the renovation radically modified Hagenbeck's project. Hagenbeck's zoo was a sort of amusement park, in which a limited number of species were chosen to offer visitors what had to appear to them as the natural experience of the wildlife. But now the zoo was a public institution, therefore its mission also changed, turning to education and knowledge. It comes, therefore, as no surprise that the administrators were carefree about the financial viability of the whole endeavour, as the zoo was intended as an educational entity.[48]

The report also stressed the colonial nature of the future zoo, as an area was devoted to ethnographic exhibits especially from "our colonies," in order to elicit the interest of the audience. Furthermore, it suggested that expenses could be reduced if the abroad diplomatic missions and the colonial administrations acquired specimens, therefore avoiding intermediation fees.[49] The press often stressed the relations between the colonial expansion and the zoo. In 1931 the Emperor of Ethiopia donated 3 lions to Mussolini and to prince Umberto, who gave the animals to the Rome zoological garden.[50] Mussolini also approved Suardi's plan to create a Zoological Civic Museum, in a joint venture with the University of Rome.[51]

The renovation of the zoo was also part of Mussolini's urban plans for Rome. As already recalled, Mussolini thought that Fascism had to deeply transform anthropologically the Italians. Such a process was not only rooted in the modification of the State and its apparatuses, but it entailed an in-depth transformation of the social and cultural relations of the citizens. It also involved a wide transformation of the urban space, especially in the Capital city.[52] As early as the 1930s, both economic and political reasons led the Fascist government to massive investments in the construction of buildings and infrastructures, mainly in Rome. The Fascist government reacted to the 1929 stock markets crash by supporting the failed banks and industrial activities, which were absorbed in a newly created institute, the IRI, *Istituto per la Ricostruzione Industriale* (Institute for the Industrial Reconstruction), and by financing a massive plan of investments in infrastructures.[53] At the same time, Mussolini boosted a profound transformation of the urban spaces of Rome, the Capital City of the new Fascist Italy. In 1932, the tenth anniversary of the Fascist Revolution, several new buildings and infrastructures were inaugurated, including the *via dell'Impero* (Empire Avenue), a large road connecting Piazza Venezia (where Mussolini had his headquarters and the site of the mass gatherings and public speeches) to the Coliseum. The new road was meant to be a monumental theatre for the military parades of the reborn Imperial Rome, but it also had – according to Gentile – a diminutive meaning for the Quirinale Palace, the official residence of the King: the new urban organization put Piazza Venezia at the center of the political life of the country.[54] A number of architects came to Rome to participate in the edification of the New Rome. Some of them were, indeed, attracted by the advantages offered by the huge state-sponsored urban projects, while others were also intrigued by the totalitarian ideology. Nonetheless, although the Regime mostly favored Marcello Piacentini's monumental design, other architects with different styles participated in the construction of buildings and infrastructures, and Pierluigi Nervi's rationalism coexisted with Piacentini's monumentalism.[55] As already stressed, Fascism was a composite phenomenon. While it can be considered totalitarian, the co-existence with other institutions and social actors left some limited space for cultural and social pluralism and, therefore, was constantly seeking consensus. There were Fascist artists, writers, philosophers, architects, but there was not a uniform Fascist art, literature, architecture or philosophy. The expansion of the zoo mirrored the same apparently chaotic pattern. Most of the new cages for animals were designed to be exotic, referring to the supposed land of origin of the animals (a tucul for the Antelopes house, African totem-style columns for the Giraffes' house and so on). However, a rationalist design characterized other buildings, such as the reptile house and the space inhabited by apes.[56] While exotic animals could be associated with human buildings, the spaces for Italian and European species tended however to reproduce the natural environment in which they lived.[57] The public was again brought to associate exotic civilisations with savage life, but to distinguish Western civilisation from it.

In October 1932, a grand opening took place. The building previously occupied by the zoo's restaurant, after a heavy renovation, became the site of the newly created Museo Civico di Zoologia (Zoology Civic Museum) and the Museo Coloniale (Colonial Museum), which moved there since its previous location was too small. The Zoology Civic Museum hosted several collections donated by the Royal Family as well as by scientists and militaries, most of them previously held in the Zoological Museum of the University of Rome. Mussolini himself officially inaugurated the two museums, and according to the reports by the *Corriere della Sera*, the Zoology Civic Museum was going to focus on scientific education, while the other establishment would be a center for colonial culture, including material aspects such as agriculture and fishing.[58] Writing to Suardi, the director of the zoo, De Bono, then Minister of the Colonies, stressed that it was decided to establish the Colonial Museum in the former restaurant of the zoo because, although two different institutions, the Colonial Museum and the zoo, represented a whole project incorporating scientific education and colonial propaganda.[59] The spatial continuity mirrored the ideal continuity between the zoo, science, and the colonies within the Fascist ideology. Suardi enthusiastically approved the project.[60] Throughout the 1930s, Fascism heavily invested in the zoo. Animals' collections grew year by year thanks to donations coming from Mussolini himself, from Italo Balbo's expedition to South America and from another regime "Gerarchi" (the hierarchs in the Fascist Party), who either were or travelled in the colonies. New specimens were traded with other zoos and acquired from Hagenbeck's firm and from other traders. Consequently, new spaces were created and old spaces were renovated. The ornithologist Jean Delacour and Alula Taibel were called to collaborate in the designing of the spaces of the zoo.[61]

Before and after the invasion of Ethiopia and the creation of the Empire in 1936, a number of scientific expeditions were sent to explore territories that the Italians poorly knew. Such expeditions often brought back animals for the zoo.[62]

A collaboration between the Colonial Museum and the zoo was established and in 1934 the latter co-financed the expedition of Saverio Patrizi to Somalia.[63] It was within this context that the Rome Zoo gained an international reputation as a scientific and educational institution. In 1938 it was decided that the congress of the International Union of the Directors of the Zoological Gardens would take place in Rome in October of the following year.[64] The beginning of WWII in September 1939 eventually led to the cancellation of the congress.[65] The war interrupted the international collaborations, although the zoo did not suffer especially from the war restrictions. In 1942, visitors even slightly increased with regards to the previous year.[66]

The Post-war Years of the Zoo and the Legacy of Italian Colonialism

By the end of WWII, the Rome Zoo was the only zoological garden in Italy, and its function and mission were unclear.[67] Italy had lost its colonies (although the protectorate over Somalia lasted until 1960) and the relation between the zoo and colonialism changed. The colonies were not any longer the main source for the new animals of the zoo, though still in 1961 the President of Somalia donated two lion cubs.[68] By the end of the war, researchers mostly reoriented their international scientific focus from the countries towards which Italy had deployed its colonial expansion (North and East Africa) to those towards which it had deployed its emigration (North and South America). It was only by 1966 that some zoologists from the University of Florence began doing research in Somalia. Their work was not connected with the Zoo.[69]

The Italian post-war centrist governments encouraged international scientific collaboration with Western countries but also with developing countries, as it was often carried out under the aegis of the international and intra-European organizations created after the war. Such collaboration was intended as a way of firmly anchoring the country inside the Western bloc.[70] Within this context, several scientific research programs involved the zoo. It was the parasitologist Ettore Biocca and the ethologist Elisabetta Visalberghi who likely carried out the most interesting research.[71] Biocca's research in a way illustrates the transition from the pre-war colonial to a post-colonial setting of the Italian scientific research system. In 1939 he was sent to the Budapest Health Institute to study filtrable viruses, with the aim of exploring the possibility of creating an institute to study polio in the Italian colonies in East Africa.[72] Soon after the war, in the early 1950s, in collaboration with the zoo staff Biocca studied parasites in vertebrates (a common practice in the history of zoo-keeping) coming from the former Italian colonies of East Africa. A decade later, between November 1962 and July 1963, Biocca organized a multidisciplinary ethno-biological expedition to study the *Yanomámi* in Amazonia. Two biologists of the Rome Zoo staff, Francesco Baschieri and Guglielmo Mangili, joined the expedition and brought back a number of Amazonian specimens for the zoo.[73]

The zoo, however, lost the centrality it had acquired during Fascism with regards to the city and within the colonial policy of the regime, having to readjust to a new and more local dimension. Throughout the 1950s and the 1960s, the administration of the zoo tried to reach the public by making famous actors and actresses patronizing animals.[74] Although the zoo increased its visitors, its budget was still in deficit at the beginning of the 1960s.[75]

Conclusion: Dealing with the Colonial Past?

Since the 1970s, the Rome Zoo did neither receive enough financial assistance nor any other kind of attention from the authorities. The structures deteriorated and animals suffered. To the visitors, the zoo looked like the animal prison described in *Зоопарк Ночю*, *Night Zoo*, a poem by Boris Slutsky: a symbol of senseless oppression. In the 1990s most of the Italian zoos were shut down and animals were sent away to more humane establishments.[76] Rome Zoological Garden was renamed Bioparco with the aim to put an end to the "animal prison." In the early 1990s the center-left mayor of Rome, Francesco Rutelli, elaborated a renovation plan to transform it into a center for the conservation of Mediterranean biodiversity. The plan actually never materialized. The Bioparco later joined an international network aiming at saving species threatened by extinction and today this is its official mission. It is worth noting how nowadays the colonial dimension of Rome Zoological Garden is not perceived by the public.

Other zoological gardens went through the same "invisibilization" of their colonial pasts up to recent times. The Barcelona Zoo is an example: in 1966, it acquired an albino gorilla from the zoological station in Ikunde (Spanish Guinea). "Copito de nieve" (Snowflake) quickly became the symbol of the zoo and a mascot for the city. Furthermore, the gorilla costs were supported by a cocoa firm operating in the Spanish colonies. Yet, the colonial meaning of Copito de nieve, conveyed by its color (a "civilized" white, as compared to "wild" black gorillas) and by its colonial origin, has never been an issue within the debate over the zoo. Colonialism was finally taken into account in 2019, when under the auspices of the mayor Ada Colau, the zoo redesigned its future strategy, focused on the conservation of "autochthonous, Iberian and Mediterranean species." The speaker of the association Zoo Twenty-first century declared: "we don't want a colonialist zoo anymore."[77]

By contrast, in Rome there has been no comparable public debate concerning the future of the Bioparco, and the role of the Rome Zoo as an instrument of colonial propaganda is almost completely removed, ultimately mirroring Italy's inability – as a former colonial power – to deal with its colonial past.[78]

Notes

1   Eric J. Hobsbawm, *The Age of Empire* (New York: Vintage Books, 1987); Eric J. Hobsbawm, *Nations and Nationalism since 1780* (Cambridge: Cambridge University Press, 1990); Christopher Bayly, *The Birth of the Modern World* (Oxford: Blackwell, 2004); Jürgen Osterhammel, *Die Verwandlung der Welt: eine Geschichte des 19. Jahrhunderts*, (München: Beck, 2010) (trans. into English as *The Transformation of the World: A Global History of the Nineteenth Century* (Princeton, NJ: Princeton University Press, 2015); Oliver Hochadel, "Watching Exotic Animals Next Door: "'Scientific' Observations at the Zoo (ca. 1870–1910)"," *Science in Context* 24, no. 2 (2011): 183–214.

2 Bayly, *The Birth of the Modern World*, 12.

3 Ibid.

4 Osterhammel, *The Transformation of the World*, 5–7.

5 On the history of the zoo, see Nigel Rothfels, *Savages and Beasts. The Birth of the Modern Zoo* (Baltimore, MD: Johns Hopkins University Press, 2002); Helen Cowie. *Exhibiting Animals in Nineteenth-Century Britain: Empathy, Education, Entertainment* (London: Palgrave Macmillan, 2014); Ito Takashi, *London Zoo and the Victorians, 1828–1859* (Woodbridge: Boydell, 2014); Mitchell G. Ash, ed., *Mensch, Tier und Zoo. Der Tiergarten Schönbrunn im internationalen Vergleich vom 18. Jahrhundert bis heute* (Wien: Böhlau, 2008).

6 Donna C. Mehos, *Science and Culture for Members only: The Amsterdam Zoo Artis in the Nineteenth Century* (Amsterdam: Amsterdam University Press, 2006).

7 Bernardin de Saint-Pierre, *Mémoire sur la nécessité de joindre une ménagerie au Jardin national des plantes de Paris* (Paris: P E Didot, 1792), 12–13.

8 Paul H. Barrett, Peter J. Gautrey, Sandra Herbert, David Kohn, and Sydney Smith, eds., *Charles Darwin's Notebooks, 1836–1844. Geology, Transmutation of Species, Metaphysical Enquiries* (Cambridge: Cambridge University Press, 1987); Oliver Hochadel. "Darwin in the Monkey Cage: The Zoological Garden as a Medium of Evolutionary Theory," in Dorothee Brantz, ed., *Beastly Natures: Animals, Humans, and the Study of History* (Charlottesville: University of Virginia Press, 2010), 81–107; John van Wyhe and Peter C. Kjærgaard, "Going the Whole Orang: Darwin, Wallace and the Natural History of Orangutans," *Studies in History and Philosophy of Science Part C: Studies in History and Philosophy of Biological and Biomedical Sciences*, 51 (2015): 53–63.

9 Mehos, *Science and Culture for Members only*, 16.

10 Gary Bruce, *Through the Lion's Gate: A History of the Berlin Zoo* (Oxford: Oxford University Press, 2017), 30; Rothfels, *Savages and Beasts*.

11 Charles Darwin, *The Expression of Emotion in Man and Animals* (Chicago, IL: Chicago University Press, 1965).

12 Rothfels, *Savages and Beasts*.

13 Laura Francescangeli, "Il Giardino Zoologico e il suo archivio: lo svago e la cultura dalla Roma di Nathan al Governatorato," in by Liliana Di Ruscio and Laura Francescangeli, eds., *I luoghi della scienza e della tecnica. Centri di ricerca musei scientifici applicazioni tecnologiche* (Rome: 2004), 49–81.

14 Nicodemo Severi, "Un Giardino Zoologico in Roma," in *La Villa ed il Giardino* 5, no. 4 (1908): 49–64.

15 See Bruce, *Through the Lion's Gate*, 100.

16 See Bruce, *Through the Lion's Gate*; Edward Said, *Orientalism* (New York: Pantheon, 1978).

17 Knottnerus-Meyer, *Guida Ufficiale del Giardino zoologico di Roma* (Genoa: Fratelli Armanino, 1911), 6.

18 Rothfels, *Savages and Beasts*, 32–33.

19 Knottnerus-Meyer, *Guida Ufficiale del Giardino zoologico di Roma*, 7.

20 Anonymous, *Perché ha fallito il giardino zoologico?* (Arpino: Società tipografica arpinate, 1913).

21 *Il Corriere della Sera*, 2 July 1928.

22 *Il Corriere della Sera*, 29 July 1932. The development of periodical state of aggressivity in Asian male elephants (rarely also in female elephants) had been observed long before. In 1879, George P. Sanderson reported that in India it was called "must" (nowadays more often spelled as musth). See G. P. Sanderson, *Wild Beasts of India. Their Haunts and Habits from Personal Observation; with an Account of the modes of capturing and taming elephants*, (London: Allen & Co., 1879), 59. However, only much later it was acknowledged that African elephants too could suffer from musth. See Joyce H. Poole and Cynthia J. Moss, "Musth in the African Elephant," *Nature* 292 (1981): 831.

23 Bruce, *Through the Lion's Gate*.

24 Guido Abbattista, *Umanità in mostra: esposizioni etniche e invenzioni esotiche in Italia (1880–1940)* (Trieste: Edizioni Università di Trieste, 2013).

25 Nicodemo Severi, "Un Giardino Zoologico in Roma," in *La Villa ed il Giardino* 5, no. 4 (1908): 49–64.

26 The Centurione and the Salvago Raggi families were closely related, sharing a common origin in the Genoan noble *élite*.

27 C. Ghezzi, *L'Istituto Coloniale Italiano e le società geografiche tra esplorazione e colonialismo, in Colonie Africane e Cultura Italiana fra Ottocento e Novecento. Le esplorazioni e la geografia* (Roma: CISU, 1995), 33–47.

28 Laura Francescangeli, *Il Giardino Zoologico e il suo archivio*; Spartaco Gippoliti, *La Giungla di Villa Borghese. I Cento Anni del Giardino Zoologico di Roma* (Rome: Edizioni Belvedere, 2010).

29 Knottnerus-Meyer, *Guida Ufficiale*.

30 Società Anonima Italiana per l'impianto e l'esercizio del Giardino zoologico di Roma. *III esercizio Assemblea generale ordinaria degli Azionisti idel 30 Marzo 1912, Relazione del consiglio d'amministrazione e dei sindaci, Bilancio al 31 dicembre 1911* (Rome: Tipografia Editrice Italia).

31 Anonymous, *Perché ha fallito il giardino zoologico?*

32 Theodor Knottnerus-Meyer, "Relazione tecnica sul Giardino Zoologico di Roma per il 1913," *Bollettino della Società Zoologica Italiana* 3, no. 1–4 (1914): 1–21.

33 Società Anonima Italiana per l'impianto e l'esercizio del Giardino zoologico di Roma. *III esercizio Assemblea generale ordinaria degli Azionisti idel 31 Marzo 1913, Relazione del consiglio d'amministrazione e dei sindaci, Bilancio al 31 dicembre 1912* (Rome: Tipografia Editrice Italia); Anonymous, "Il concordato preventivo della Società del 'Giardino Zoologico'", *Il Corriere della Sera*, 22 August 1912, 4.

34 Abbattista, *Umanità in Mostra*, especially on chapter 9; Guido Abbattista and Nicola Labanca, "Living Ethnological and Colonial Exhibitions in Liberal and Fascist Italy," in P. Blanchard et al., eds., *Human Zoos: Science and Spectacle in the Age of Colonial Empires* (Liverpool: Liverpool University Press, 2008), 341–352.

35 Società Anonima Italiana per l'impianto e l'esercizio del Giardino zoologico di Roma. *III esercizio Assemblea generale ordinaria degli Azionisti del 28 Marzo 1914, Relazione del consiglio d'amministrazione e dei sindaci, Bilancio al 31 dicembre 1913* (Rome: Tipografia Editrice Italia).

36 Bruce, *Through the Lion's Gate*; Alessandro Ghigi, "Il giardino zoologico di Roma," *Il Corriere della Sera*, 10 August 1912, 3.

37 The Italian Zoological Society was not the main disciplinary association in the field and was eventually discontinued together with its own journal in 1919. The Unione Zoologica Italiana (Zoological Italian Union) founded in 1900 was the society that gathered Italian zoologists.

38 Osterhammel, *The Transformation of the World*, 263.

39 Nicola Labanca, *Oltremare. Storia dell'espansione coloniale* (Bologna: il Mulino, 2002); Alberto Aquarone, "Politica estera e organizzazione del consenso nell'età giolittiana: il congresso dell'Asmara e la fondazione dell'Istituto coloniale italiano," *Storia Contemporanea* 8, no. 1-2-3 (1977): 57–119, 291–334, 549–570.

40 Sabino Cassese, *Lo Stato fascista* (Bologna: il Mulino, 2010). Furthermore, Claudio Pavone stressed that the continuity between the fascist and the democratic state was not just matter of laws, but it was deeply rooted in the civil servant's agency. The judges, for instance, applied the norms on epuration in such a way that most of the former Fascists were freed by any charge. Claudio Pavone, "La continuità dello stato. Istituzioni e uomini," in C. Pavone, ed., *Gli uomini e la storia. Partecipazione e disinteresse nella storia d'Italia* (Turin: Boringhieri, 2020). See also H. Woller, *Die Abrechnung mit dem Faschismus in Italien 1943 bis 1948* (Munich: Oldenburg, 1996). On the Fascist state see also Leonida Tedoldi, *Storia dello Stato Italiano* (Rome-Bari: Laterza, 2018).

41 Cassese, *Lo Stato fascista*; Emilio Gentile, *La via italiana al totalitarismo* (Rome: Carocci, 2008). See also Hannah Arendt, *The origins of Totalitarianism* (Berlin: Shocken Books, 1951); Carl J. Friederich and Zbigniew K. Brzezinsky, *Totalitarian Dictatorship and Autocracy* (Cambridge: Harvard University Press, 1965); Alberto Aquarone, *L'organizzazione dello stato totalitario* (Turin: Einaudi, 1965); Renzo De Felice, *Mussolini 8 vols* (Turin: Einaudi, 1965), 97.

42 *Il Corriere della Sera*, 4 March 1924.

43 Gippoliti, *La Giungla di Villa Borghese*, 87.

44 Letter from Suardi to the Governor of Rome, 3 September 1940; Letter from Suardi to the Governor of Rome, 1 April 1941 in ASC, Segr. Generale 1923-48, Carteggio, busta 151.

45 *Il Corriere della Sera*, 22 January 1928.

46 Gippoliti, *La Giungla di Villa Borghese*, 86.

47 SPQR Governatorato di Roma, *Relazione sull'esercizio 1929 VIII* (Rome: Tipografia Cuggiani, 1930).

48 Ibid.

49 Letter from the Azienda Giardino Zoologico del Governatorato di Roma to S.E. il Cav. Benito Mussolini Primo Ministro Segretario di Stato Per l'Interno 29 November 1931. In Central State Archive, Rome, Particular Secretary of il Duce, Rome (henceforth ACS/SPD), Busta 510, fasc. 197.032.

50 *Il Corriere della Sera*, 6 May 1931.

51 Appunto, 14 May 1932 in ACS/SPD, Busta 510, fasc. 197.032.

52 See Antonio Cederna, *Mussolini urbanista. Lo sventramento di Roma negli anni del consenso* (Rome-Bari: Laterza, 1979); Italo Insolera, *Roma moderna* (Turin: Einaudi, 2001); Emilio Gentile, *Fascismo di Pietra* (Rome-Bari: Laterza, 2007).

53 Rolf Petri, *Storia economica d'Italia. Dalla Grande guerra al miracolo economico (1918–1963)* (Bologna: Il Mulino, 2002).

54 Gentile, *Fascismo di Pietra*, 93–94.

55 Ibid., 96.

56 Gippoliti, *La Giungla di Villa Borghese*, 86.

57 Ibid., 95.

58 *Il Corriere della Sera*, 7 November 1932.

59 Letter from De Bono to Suardi of 20 January 1932; Letter from De Bono to Boncompagni, Governor of Rome of 20 January 1932 in ASC, Segr. Generale, Carteggio 1923-1948, Busta 151.

60 Letter from Suardi to De Bono of 5 February 1932 in ASC, Segr. Generale, Carteggio 1923-1948, Busta 151.

61 Gippoliti, *La Giungla di Villa Borghese*, 92–96.

62 Ibid.; Francesco Surdich "Le spedizioni scientifiche in Africa Orientale," in Angelo del Boca, ed., *Le guerre coloniali del Fascismo* (Rome-Bari: Laterza, 2008); Roberto Maiocchi "Italian scientists and the war in Ethiopia," *Rendiconti dell'Accademia Nazionale delle Scienze detta dei XL Memorie di Scienze Fisiche e Naturali* 133° (2015), Vol. XXXIX, Parte II, Tomo I: 127–146.

63 Leonardo Lautella "Le spedizioni scientifiche di Saverio Patrizi, entomologo, biospeleologo e cacciatore," *Memorie del Museo Civico Di Storia Naturale Di Verona – 2a Serie. Monografie Naturalistiche* 4 (2019): 53–57.

64 "Da una lettera dell'Ufficio CIT di Amsterdam"; Governatorato di Roma, Servizio Giardino Zoologico, Unione Direttori Giardini Zoologici, Conferenza 1939, XVII in ASC, Arch. Giardino Zoologico, Carteggio, Busta 436.

65 Letter from Lamberto Crudi, 22 September 1939 in ASC, Arch. Giardino Zoologico, Carteggio, Busta 436.

66 Letter from Crudi to Suardi, 23 August 1942; Letter from Crudi to Suardi, 19 October 1942 in ASC, Arch. Giardino Zoologico, Carteggio, Busta 629.

67 *Il Corriere della Sera*, 28 August 1945.

68 Il Tempo, 26 July 1961.
69 Antonio Ercolini, "Leo Pardi e il centro di studio per la faunistica ed ecologia tropicali del consiglio nazionale delle ricerche dal 1971 al 1985," *Monitore Zoologico Italiano. Supplemento* 22, no. 1 (1987): 13–20.
70 On scientific policy in post-war Italy, see Giovanni Berlinguer, *Politica della scienza* (Rome: Editori Riuniti, 1970); Antonio Ruberti, "Riflessioni sul sistema della ricerca dopo il 1945," in Raffaella Simili, ed., *Ricerca e istituzioni scientifiche in Italia* (Rome-Bari: Laterza, 1998), 213–230; Giuliana Gemelli, "Epsilon effects: biomedical research in Italy between institutional backwardness and islands of innovation (1920s–1960s)," in Giuliana Gemelli, Jean-François Picard, and William H. Schneider, eds., *Managing Medical Research in Europe: The Role of the Rockefeller Foundation (1920s–1950s)* (Bologna: CLUEB, 1999), 173–197; Mauro Capocci, "Politiche e istituzioni della scienza: dalla ricostruzione alla crisi," in Claudio Pogliano and Francesco Cassata, eds., *Storia d'Italia. Annali 26. Scienza e cultura dell'Italia unita* (Turin: Einaudi, 2011), 265–294.
71 On Visalberghi's research see Elisabetta Visalberghi, *Idee per diventare etologo. Osservare il comportamento degli animali* (Bologna: Zanichelli, 2006). On Biocca see Ettore Biocca, *Viaggi tra gli Indi* (Rome: CNR, 1965); M. Coluzzi, "Ettore Biocca. Un irraggiungibile esempio di vita," *Parassitologia* 44 (2002): 1–4; Daniele Cozzoli, "Ethobiology during the Cold War: Biocca's Expedition to Amazonia," *Centaurus* 58 (2016): 281–389.
72 Letter from the Director of the National Research Council to the Direzione di Sanità Militare Roma of 19 April 1939 in ACS/CNR/Comitato di Biologia e Medicina, busta 16, fasc. 54. On polio in Hungary See Dóra Vargha, *Polio Across the Iron Curtain: Hungary's Cold War with an Epidemic* (Cambridge: Cambridge University Press, 2021).
73 On Biocca's expedition see Biocca, *Viaggi tra gli Indi I*; Cozzoli, *Ethobiology during the Cold War.*
74 "Due attrici gemelle svedesi madrine di due orsi allo zoo," press release of 7 April 1961; "Elsa Martinelli imporrà il nome a due leopardi nati allo zoo," press release of 4 July 1961 in AS, Arch. Giardino Zoologico, Carteggio, Busta 629.
75 "Opere per 650 milioni al Giardino Zoologico di Roma" in *Il Globo*, 5 March 1961.
76 Gippoliti, *La Giungla di Villa Borghese*, 11.
77 *El Periodico*, 3 May 2019, retrieved on 28 February 2020 https://www.elperiodico.com/es/barcelona/20190503/aprobada-reforma-zoo-barcelona-7436693.
78 The authors are particularly indebted with Oliver Hochadel with whom not only we discussed several aspects of the zoo in the nineteenth and twentieth century, but who also carefully read this paper and made invaluable comments. We also wish to thank Spartaco Gippoliti (for providing information and the invaluable help in accessing relevant printed sources) and Daniela Ronzitti (Archivio Storico Capitolino, Rome) for the help in locating and accessing archival sources.

Part 2
Knowledge Production at Imperial Crossroads

8 The Astronomical Observations of Bento Sanches Dorta in Rio de Janeiro, 1781–1787

Heloisa Meireles Gesteira

Introduction

On or around August 6, 1787, the British First Fleet entered the Bay of Guanabara, near the city of Rio de Janeiro, the capital of the Portuguese colony of Brazil. Aboard one of the ships that made up the fleet, the HMS *Sirius*, was one William Dawes, a lieutenant in the British Navy who went on to take part in the establishment of the colony of New South Wales, Australia. In addition to being a colonial administrator and surveyor, Dawes would establish an observatory in the new colony, on what is presently known as Dawes Point, in Sydney. This mission was assigned to him by Nevil Maskelyne, Astronomer Royal at Greenwich at the time.[1]

The First Fleet ships were stationed in the Bay of Guanabara near the island of Enxadas until September 3, when they set sail towards the Cape of Good Hope to reach their destination in Australia. During the time that the fleet was anchored in the Bay of Guanabara, the crew, who were not in principle allowed to disembark nor enter the city of Rio de Janeiro, took their time to repair the vessels, replenish supplies, and set up a station to carry out astronomical observations during this period. In a letter addressed by Dawes to Maskelyne, we learn that Dawes met the Portuguese astronomer Bento Sanches Dorta, first on the island of Enxadas and later in the city, at Sanches Dorta's workplace.[2]

Both men were brought to Rio de Janeiro by the value put in their skills as astronomers by their respective imperial governments. Sanches Dorta had been commissioned years before to help settle border disputes between Portugal and Spain in South America, while Dawes was on the *Sirius* with instructions to carry out observations and tests intended to perfect proposed solutions to the problem of establishing longitudes at sea. Astronomical observations for the determination of the coordinates of places on land, as well as for the orientation of moving ships along maritime routes, required men capable of using instruments, registering the data, and making the necessary calculations. Both navigation and the ability to accurately pinpoint the location of cities and islands were matters of obvious importance to the administration of overseas empires. Moreover, any solution to the problem of longitude at sea, including John Harrison's famous maritime chronometer, necessarily had to be tested against the much more reliable data coming from land-based astronomical observations.[3] What Dawes found in Rio de Janeiro was

DOI: 10.4324/9780367814540-10

precisely the kind of data he needed to check his methods and the Harrison piece he was carrying. Indeed, over many years of laborious observation and calculations, Sanches Dorta had reached highly accurate values for the place's longitude in relation to Lisbon, Paris, and Greenwich. Reciprocally, the Portuguese found in Dawes and his instruments an opportunity to accrue prestige and have his skills recognized. As I argue in this essay, instruments and the material practices of observation were at the center of the exchanges that took place between those two men who were put in front of each other as a direct result of the entanglement of science and empires.

Rio de Janeiro, Imperial Hub of the South Atlantic

In 1740, the Jesuit Diogo Soares gave Rio de Janeiro a very special place in his map *Nova e primeira carta da Terra firme, e costas do Brasil ao Meridiano do Rio de Janeiro, desde o Rio da Prata athe o Cabo Frio com novo caminho do (...) Rio Grande athe a cidade de São Paulo, ao poderosíssimo Rei D. João.*[4] Soares, who had travelled extensively across Portuguese America, chose Rio de Janeiro as the site of the prime meridian. This choice has been interpreted by Portuguese historian Jaime Cortesão as the way that Soares, royal geographer nominated by king D. João V to map Portugal's American possessions, found to prevent the geographic information contained in his maps from being intelligible by rival powers, since the city's longitude was poorly known.[5] However, the emphasis given to Rio de Janeiro can also be interpreted as a reinforcement of its centrality, both for the colonial administration and for imperial interests at that time.

The strategic value of the location of Rio de Janeiro had been debated at the court in Lisbon since the beginning of the eighteenth century, with one of its promoters being D. Luís da Cunha, an ambassador who had spent time in influential political centers of Europe – London, the Hague, Paris, Brussels, and Madrid – and participated in important negotiations, such as the Treaty of Utrecht (1713–1715). Through his involvement in European courts, D. Luís da Cunha was a defender of Portuguese territorial interests in the Americas. At the end of his life, he wrote his *Political Instructions*, where he suggested that the Portuguese monarchy be transferred to the Americas, and more precisely to Rio de Janeiro.[6]

In the second half of the seventeenth century, Rio de Janeiro started to establish itself as the main colonial city in the South Atlantic. Competing with Salvador, it asserted its importance for the administration of the southern Portuguese possessions. During the Iberian Union (1580–1640), the permission to bring enslaved Africans to forced labor in the Potosí mines was granted by the court in Madrid to Portuguese colonists in Rio de Janeiro. This meant that the South American city became a strategic port of entry for the enslaving routes that crossed the South Atlantic from Luanda. Enslaved Africans were then sent from Rio de Janeiro to the Rio de la Plata basin, and thence to the mining region of Potosí.

The actions of Governor Salvador Correia de Sá e Benevides helped to strengthen the position of Rio de Janeiro in the Portuguese imperial balance. When he received news of the dissolution of the Iberian Union in 1640, with the acclamation of D. João IV as king of Portugal, Correia de Sá immediately swore loyalty to the new king, carrying with him the southern territories of the colony, over which Rio de Janeiro exercised administrative control. Two years later, Correia de Sá was appointed to the Royal Overseas Council, where he actively defended the interests of Rio de Janeiro in imperial policies. When he returned to the position of governor of the city of São Sebastião, Correia de Sá carried with him the title of "Governor and Captain General of the Southern Division," contributing to the centrality of Rio de Janeiro along the coast and extended inland areas connected to it, from Espírito Santo to the Plata basin. Both he and his family were actively involved in the Atlantic slave trade and he played an important role in the wars which regained Angola for the Portuguese in 1648.[7]

The roads opened between Rio de Janeiro and the mining regions further inland at the end of the seventeenth and the beginning of the eighteenth centuries reinforced the position of the city as the center of the southern Portuguese territories in America. The discovery of another gold-producing region in the hinterlands further strengthened the centrality of Rio de Janeiro. A decisive moment for this involved the action of another governor of the city, Gomes Freire de Andrade, in office from 1733 to 1763. When Gomes Freire was appointed viceroy of Brazil in 1763, he decided to stay in the city, instead of moving to Salvador, the nominal capital of the colony. From then on, all viceroys took residence in Rio de Janeiro, which became the *de facto* center of political power.[8]

Furthermore, the location of Rio de Janeiro placed the city on the path of the maritime currents of the South Atlantic, and for this reason, it was an obligatory port of call for the ships going to India, either through the Strait of Magellan or along the Cape route. Of interest here are the reflections of Richard Drayton, who argued that the maritime currents were important agents in the shaping of overseas networks, stimulating connections between places and individuals and initiating the flow of goods. According to Drayton, it is important to understand how "natural facts, written into the structure of the seas, shaped how the British, but not limited to the British, amassed both knowledge and imperial power."[9]

It should be noted that the same maritime currents that shaped imperial routes allowed for autonomous connections between astronomers on board ships. Even though these men might be at the service of different empires, they could take advantage of chance meetings that took place because of forced stops. This could sometimes result in personal advantages and certainly promoted a type of exchange of ideas that eluded imperial control. Thus Rio de Janeiro, where numerous ships could not avoid calling port, cannot be viewed through the traditional prism of the cultural isolation supposedly inherent to its colonial condition. In reality, the late-eighteenth-century intellectual milieu of the city was quite vibrant. The city boasted of an Academy of Sciences (1772–1779), a botanical garden, a Literary Society (1786–1794), and a natural history cabinet (Casa dos Pássaros), created in 1784 with encouragement and support from the viceroy D. Luís de Vasconcellos. Between 1781 and 1788, a sheer number of astronomical and meteorological observations were carried out by Bento Sanches Dorta, while he lived in the city.

Astronomical and Meteorological Observations in Rio de Janeiro (1781–1788)

Between 1781 and 1788, Sanches Dorta carried out astronomical and physical observations in the city of Rio de Janeiro. His presence in the city came as a result of the Treaty of St. Ildefonso, signed by Portugal and Spain in 1777. The treaty intended to settle border disputes between the two empires in South America. Sanches Dorta was part of the team of astronomers and mathematicians commissioned by the Portuguese crown to carry out the fieldwork stipulated by the treaty. Sanches Dorta and another Portuguese astronomer, Francisco de Oliveira Barboza, were assigned to make determinations of geographical coordinates in the region of São Paulo, south of Rio de Janeiro; other teams were to work further south. However, Sanches Dorta and Oliveira Barboza remained in the city of Rio de Janeiro for seven years and only after that did they finally move to São Paulo, where they made astronomical and meteorological observations and surveyed the coastline.

Sanches Dorta was born in Coimbra in 1739. The son of a goldsmith, he practiced the craft briefly, but, in his early thirties, decided to enrol in the University of Coimbra, in the newly created Mathematics and Natural Philosophy courses. After completing his studies, he applied to join the engineering corps of the Portuguese military, and then, along with several other Coimbra-trained mathematicians, he was commissioned to carry on the border-settling fieldwork in the wake of the Treaty of St. Ildefonso. During the time he lived in Rio de Janeiro, Sanches Dorta became a corresponding fellow of the Royal Academy of Sciences in Lisbon and was later promoted to full fellow. Part of his scientific work was published in the Journal of the Academy. In 1788, he went to São Paulo, where he would die in 1795.[10]

It is important to highlight the working conditions that Sanches Dorta created for performing his observations in Rio de Janeiro. Armed with a respectable collection of instruments and books, he improvised a small observatory. The quality of his observations and the existence of a well-equipped space certainly facilitated his interactions with William Dawes, even if they lasted for a very short time. Analyzing the design and placement of the observation station and the instruments employed is important for our study since both the organization of the workplace and the equipment in use were crucial clues for Sanches Dorta and Dawes to recognize, and ultimately respect (or not) each other, despite their governments' rivalries.

During his stay in Rio de Janeiro, Sanches Dorta organized, in his residence near Morro do Castelo (Castle Hill), a space where he could perform astronomical observations every day. The practice of keeping daily records of natural phenomena was consolidated throughout the eighteenth century and was driven, among other factors, by the increasingly frequent use of precision instruments, i.e., ones that allowed the recording of data as numbers. Observations with instruments were carried out not only within dedicated institutions but also in domestic and other settings.

Whether for scientific reasons or otherwise, the fact is that the daily operation of merchant ships and others involved in colonial missions increasingly employed optical and mathematical instruments, such as telescopes, astronomical quadrants, thermometers, barometers, and pendulums, among others. This led to the establishment of observation stations and observatories on many colonial outposts. The instruments were important for the compilation of the physical and geographical descriptions of the territories. As this practice developed, instruments and manuals were transported along with practitioners skilled in using them on geodesic and astronomical missions, and border-settling expeditions. The results were recorded in notebooks or diaries and collected in scientific institutions located in Europe and elsewhere.

The creation of observatories linked to state interests, such as that of Paris, in 1666, and Greenwich, in 1675, in addition to increasing astronomical observations in the service of navigation, surveying, and control of imperial territories, allowed state-sponsored institutions to accumulate information, and promoted exchanges between astronomers. In Portugal, king D. João V (r. 1706–1750) supported the purchase of instruments for the observatories of the Jesuit College of St. Antão and the Royal Palace, both in Lisbon. These places, in addition to being equipped for astronomical observations, were important spaces for the training of engineers and technicians qualified to undertake surveying work in Portugal and its domains. The aforementioned Diogo Soares, together with fellow Jesuit Domenico Capacci, both associated with the College of St. Antão, were sent to Portuguese America by order of D. João V and stayed there between 1730 and 1748. They were entrusted with the task of producing an atlas of the lands belonging to Portugal. To carry the task out, they took precision instruments with them.[11]

In a detailed study on the employment of astronomical quadrants manufactured in England by observatories across Europe during the eighteenth century, Jim Bennet demonstrates how the use of like instruments facilitated the communication of the results obtained by astronomers, led to the formation of an international community, helped to create consensus in the field of astronomy, and stimulated the establishment of similar research programs.[12] Instruments drove the uniformization of observation practices, since the use of an instrument often required the adoption of standardized gestures and procedures that circulated in "user manuals" written by manufacturers or in astronomy textbooks. In addition, observation data circulated not only in the form of numerical tables; these were complemented by texts that reported on the instruments used and, in some cases, their manufacturers and specifications, all of which allowed specialized readers to understand observation routines and calculations employed to get to the results.

The wealth of details present in the writings of Sanches Dorta allows us today to reconstruct his work routines in Rio de Janeiro, even if partially. Over the years that he lived in Rio de Janeiro, Sanches Dorta carried out his work at what he called an *estação de observação* (observation station). A strong indication of his permanence in the same place is that in his writings, published in the transactions of the Royal Academy of Sciences of Lisbon, Sanches Dorta refers to Castle Hill numerous times. From there he recorded the altitude of the Sun and other stars on meridian passage, eclipses of the Moon, Sun and Jupiter's satellites, and meteorological phenomena. The care that Sanches Dorta took in recording every detail, from the instruments he employed to the procedures and routines, allows us to infer that he had made all the observations in a fixed place. Furthermore, keeping the observation point fixed is important to arrive at more accurate mean values, as in the case of geographic coordinates, and a more consistent series from the point of view of meteorological observations.

Although this was not initially planned, during the years that Sanches Dorta and Francisco de Oliveira Barboza awaited in Rio de Janeiro for the order to proceed to São Paulo, both found support to work in well-equipped observation stations in the city. We are avoiding calling these stations "observatories" because the latter may evoke special-purpose buildings that accommodate the precision instruments in the most stable way possible or conjure images of centralization of information and close-knit ties with the state, none of which was the case in Rio de Janeiro. Indeed, the stability of the precision instruments was one of the permanent challenges, especially in improvised stations and in field observations made during trips. Astronomical data collected under these circumstances usually underwent further processing after travelling along imperial pathways that led to better-equipped institutions, which were also granters of prestige for actors who managed to get into their networks. Such data could be used for determining geographical coordinates, or employed in geodetic studies and mapmaking, among other possibilities.

Many colonial observation stations appeared during the second half of the eighteenth century, such as the ones in Rio de Janeiro or Dawes's in New South Wales. The term "colonial" refers here to the kind of ties that linked these places to European metropolitan institutions, which grew increasingly dependent on data obtained in this way. The Rio de Janeiro station was connected, above all, to the Observatory of the Royal Academy of Sciences in Lisbon, an institution to which Sanches Dorta sent part of his data and from where he extracted information and prestige. It was not by chance that he was elected a member of this Academy precisely during his stay in Rio de Janeiro.

In a letter written by Secretary of State Martinho de Melo e Castro, dated February 17, 1781, addressed to D. Luís de Vasconcelos, Viceroy of Brazil, we learn of the instruments that arrived in Rio de Janeiro in that year:

> ...a quadrant with all its parts; a half-second pendulum with its parts; two achromatic lenses with the pedestal and more pieces for observations of Jupiter's satellites; a large azimuth needle with its pedestal plus attached parts; two small metal needles to determine [directions] and magnetic azimuths; a circular reflection instrument for observing distances; a pocket watch for marking the seconds; a complete mathematical kit and a nautical sextant for observing the altitudes and distances of the stars.[13]

In addition to the equipment for making drawings and repairing the instruments, Sanches Dorta reported that, "all these instruments are part of a collection that her majesty ordered be delivered to us when we left Lisbon with the task of the Demarcation of South America between Portugal and Spain."[14]

It is important to notice that the list indicated the use each instrument would be given: measuring (angular) distances, observing Jupiter's satellites, marking time.

The collection was organized and purchased in London by the Portuguese natural philosopher residing in that city, João Jacinto de Magalhães, or J. H. Magellan, as he was known in London. The purchasing of mathematical instruments for use in observatories and fieldwork was an undertaking that required considerable practical knowledge. It was not simply a matter of choosing objects from the shelves of London's many workshops filled with scientific instruments. The business of instruments for scientific use and technical work was geared towards restricted and specialized audiences. The needs as well as the social standing of customers had to be taken into consideration. In this case, Magalhães had to bear in mind that the commissioner was the Portuguese crown itself, which had a stated interest in fostering scientific enterprises in metropolitan Portugal and carrying out topographical surveys and settling borders in the American colony. The real-world conditions in which the artefacts were to be used also had to be anticipated. The collection to be sent to America was expected to be in constant movement on difficult terrain. Furthermore, the observation protocols in which the instruments would be employed had to be considered.

The care needed for the acquisition of the instruments often required that the buyer himself closely monitored the production of the artefacts, aware of the needs of those who would use the instruments and the conditions of their use. Regarding the astronomical quadrant that would be transported to the far reaches of America, the dimensions of the instrument were something that worried Magalhães:

> Portable quadrants must be 12 to 15, or at most, 23 inches in radius. They are the most appropriate instruments for Astronomy, particularly when travelling, and are used to make any kind of observation. The medium size of the instrument facilitates its transportation and, when it is well made and equipped with all the advantages provided by English astronomers and craftsmen, is the most appropriate, the most useful and the most esteemed among all instruments.[15]

Magalhães also oversaw the purchase of part of the instruments for an observatory associated with the Royal Academy of Sciences newly created in Lisbon in 1779, by D. João Carlos de Bragança, Duke of Lafões and the naturalist José Correia da Serra. Before the work of the Academy began in earnest in 1787, the instruments were already provisionally mounted at the Castle of S. Jorge, in Lisbon. There, Miguel Antonio Ciera made observations with the artefacts acquired by Magalhães. It is plausible to suppose that Sanches Dorta worked in Rio de Janeiro under conditions akin to those enjoyed by astronomers linked to the Royal Academy of Sciences in Lisbon.

Upon receiving the collection of instruments (and several books), Sanches Dorta was ready to mount a colonial observatory in Rio de Janeiro, where he would work during the seven years that he spent in the city. To reconstruct his workspace, we can resort to his published observation diaries, where Sanches Dorta documented in detail the astronomical observations undertaken between 1781 and 1787 (the meteorological observations continued into 1788). Some important information about the observatory is found only in the entries for the first year, because the astronomer considered it important to follow unchanging protocols. In respect to his meteorological observations, for instance, Sanches Dorta declares:

> It is my duty to announce that from this point onward, unless there were some change (which I would duly notice), all the instruments that I employed were placed in the chamber of my dwelling, which is located 50 feet and 4 inches above sea level. This chamber has three windows facing southwest, and I almost always keep them open. In this same chamber, I drew an exact meridian, where I constantly keep the compass to account for the declination of the needle.[16]

It was in this chamber that Sanches Dorta repeated a daily ritual of observations for many years. He would register weather data with a thermometer and a barometer (mentioned only in the observations of 1784), and magnetic variation with the compass needle.[17] For the determinations of latitude and longitude, the procedures were also recorded in the diaries, and we learn that Sanches Dorta always used the same methods and instruments:

> The meridian altitudes of the Sun, and Stars were taken with an astronomical quadrant with a one-foot radius, built by Mr. Sisson, a London artisan, in the year 1779. The eclipses of Jupiter's satellites were observed with Dollon's achromatic lenses, one with a focal distance of 3½ feet, and another of 17 inches. True time was determined through an excellent pendulum of seconds, calibrated by many corresponding solar altitudes, taken before and after the observations.[18]

Between the years 1781 and 1787, Sanches Dorta used the large Dollond achromatic refractor to perform many observations of Jupiter's satellites, in order to determine the longitude of the city. On average, he made 45 observations of the eclipses of Jupiter's moons every year. The pendulum was always adjusted the day before, the day of, and the day after the observations, to yield the true solar time. At times, the *Connoissance des Temps*, a regular publication from the Paris Observatory that contained the main astronomical events for each year, was used to calculate the time difference between his eclipse sightings and their expected times at the meridian of Paris. Sanches Dorta also used eclipse times determined in Lisbon by Miguel Antonio Ciera. Results of several independent observations were averaged to obtain the longitude as accurately as possible, as is shown in this passage of Sanches Dorta's diaries:

> Now, taking the arithmetic mean of these seven results, we have the difference of the meridians of Lisbon and Rio de Janeiro at 2h16'31.3", and in relation to the observatory of the Royal Academy of Sciences, at 2h16'35.3". It follows that the longitude of Rio de Janeiro is 34°8'50" west of the meridian of the Observatory of the Royal Academy of Sciences of Lisbon; and

45°37'50" west of the Royal Observatory of Paris. From the westernmost part of Ferro Island, the longitude of Rio de Janeiro is 34°52'10".[19]

The same kind of ritual was carried out to determine the altitude of the Sun and other stars on meridian passage, with a view to establishing the latitude of Rio de Janeiro: Sanches Dorta adjusted the pendulum and recorded the direct result of the observations, performed this time with the one-foot quadrant. Averaging the results of many Sun and stellar observations, Sanches Dorta determined the latitude of Rio de Janeiro to be 22°54'12".[20]

The diaries show that although there were minor changes, the procedures were essentially maintained over time. For the year 1787, when William Dawes visited Rio de Janeiro, Sanches Dorta reported having adopted the same procedures as in previous years, "both in the manner and time of making the observations, as well as using the same instruments (…) and keeping them in the same position."[21] The observation diary for 1787 contains, first, twelve tables that detail the meteorological observations with data taken from the barometer, thermometer, and magnetic needle. Then follows a description of the instruments and the temperature highs, lows, and averages for the year. The damage caused by a lightning storm that hit a sloop anchored in the Bay of Guanabara and the observation of a halo around the Moon are highlighted. In addition, Sanches Dorta notes that he observed the zodiacal light twelve times, and the southern lights seven times. Finally, he presents the tables with the usual astronomical observations, i.e., of the eclipses of Jupiter's satellites with the large refractor, and of meridian passages with the one-foot quadrant.

News from the Island of Enxadas

William Dawes's letter to Nevil Maskelyne describing observations performed on Enxadas, in the interior of the Bay of Guanabara, provides us with clues to think about how knowledge is constructed and circulates. In the case of Dawes's meeting with Sanches Dorta, it was the observation instruments and practices that facilitated the communication between both men. The letter tells us that on or around August 6, 1787, the English ships approached Cape Frio, on the bay's entrance, to wait for the right conditions to call port at the main harbor in Rio de Janeiro. Dawes goes on to report that on August 11, their entrance to the main port having been denied, the crew decided to set up camp on the island, and his astronomical instruments were duly deployed. The ships would end up moored for almost a month before setting sail to Australia, where Dawes would, at last, assume his official duties as a colonial administrator, astronomer, and surveyor.

The letter shows that Dawes was following instructions from the Astronomer Royal to seize every opportunity to set up a station to make astronomical observations and check the Harrison timekeeper he was carrying. During the second half of the eighteenth century, several tests with timekeepers were carried out, starting with a trip to Jamaica in 1761–2 in which the so-called H4 chronometer was first tested. Although the H4 was approved, this did not entail its immediate adoption, even on British ships. As demonstrated by Jim Bennet, the process was marked by disputes and the test of the Harrison piece ended up raising many additional questions, such as: could the apparatus be reproduced, including by other artisans? Could production be scaled up to meet naval demand? Was it more reliable than astronomically based methods, particularly lunar distances? It was by this time, precisely in 1765, that Nevil Maskelyne assumed the post of Astronomer Royal at Greenwich, where he was charged with the task of stimulating the search for the resolution of the problem of longitude at sea, and testing methods and instruments.[22]

One of the most pressing problems detected in the tests carried out in that first trip to Jamaica, long before Sanches Dorta and Dawes's chance encounter, was the serious dearth of accurate longitude data against which to test the chronometer. This would go on for decades. However, such was not the case with Rio de Janeiro, a city located at the intersection of imperial maritime pathways, the main Portuguese port in the South Atlantic, and not unknown to other astronomers who had previously made observations there.[23] And, especially through the years-long work of Sanches Dorta, by 1787, when Dawes visited the city, its longitude in relation to the meridians of Paris, Greenwich, and Ferro Island, in the Canary archipelago, was very well determined. Dawes had privileged access to the tables produced by Sanches Dorta at his station in the city center long before the Portuguese astronomer had them published.[24]

On Enxadas, Dawes, equipped with his own instruments, took the opportunity to set up an observation station. His letter to Maskelyne describes the instruments, procedures, and results obtained. The first artefact Dawes refers to is an astronomical quadrant. After carefully adjusting the instrument, he recorded equal altitudes of the Sun to determine the local time and made the first tests of the chronometer, estimating the time difference in relation to Greenwich to be 2h50'52". Then Dawes turns to the astronomical pendulum clock, whose mounting he describes in great detail, since it was his main resource for testing the timekeeper.[25] Between August 16 and 20, Dawes worked at his station, adjusting the local time and calibrating the pendulum clock, in addition to observing the occultation of Jupiter's satellites with a refractor and marking the position of the Moon along the ecliptic.

But what is most telling in this letter happened on the morning of August 17. Dawes reports the visit of two astronomers who lived in Rio de Janeiro at the service of the Portuguese crown and tells Maskelyne he "immediately foresaw that [he] might possibly get some observations from them which would be acceptable to you. & therefore determined if possible to return their visit before our quitting this place: they admired the quadrant and took notice how equal the beats of the clock were."[26]

The visitors' admiration for the instruments caught Dawes's attention, especially their sensitivity to the regularity of the clock's beats. The use of instruments required mathematical knowledge for the calculations, but, above all, practical skills for their correct manipulation. In the case of astronomical observations, precise training of the eye and ear was necessary so that the observer could capture the exact timing of what was being seen, such as the disappearance of a satellite of Jupiter. The quality of the results depended as much on the quality of the artefacts as on the training and sensitivity of the astronomer. It was necessary to assure "the painstaking coordination of eye and ear, matching star motions along the grid of an eyepiece micrometer with the beats of a nearby pendulum clock."[27] In noticing the Portuguese astronomers' immediate awareness of the quality of his clock, Dawes at once recognizes a skill shared by them and himself and takes pride in the quality of English instruments.

Dawes goes on to tell Maskelyne that he returned the visit of the Portuguese astronomers. In Rio de Janeiro, he met one of them at the observation station. As we will learn, it was Sanches Dorta, who kindly allowed the Englishman to copy some data. "I might have copied much more had time allowed it," notes Dawes.[28] In addition, Dawes reports that his host has conducted meteorological observations and observed many eclipses of Jupiter's satellites from his station in Rio de Janeiro, "any or all of which he would with pleasure communicate to you if you should think proper to enter into a correspondence with him; all that he wish'd in return was observations of Jupiter's satellites made at Greenwich."[29] Dawes notices that although the place was packed with astronomy books in French, it seemed better that Maskelyne addressed Sanches Dorta in Latin. Finally, Dawes notes that "The address of the gentlemen I found at home is as follows: Ao Senhor Bento Sanchez d'Orta no Rio de Janeiro," the house apparently being the same one in which La Caille had stayed.

Conclusion

The routes and currents of the Atlantic help us to understand why the position of some port cities, such as Rio de Janeiro, appear prominently on nautical tables from the eighteenth century. An obligatory stop for ships, the city was located on the route of ships that crossed the South Atlantic, many of which counted with the presence of astronomers on board. In the case of William Dawes, we know that Nevil Maskelyne had given him instructions in relation to the procedures for test-ing the maritime chronometer. Years before, Maskelyne himself had taken part in trials concerning the timekeeper and had been involved in the efforts to determine the longitude of Greenwich. Similarly, Sanches Dorta had been brought to Rio de Janeiro to carry on official duties as an astronomer employed by the Portuguese crown. Equipped with books and instruments, Sanches Dorta was able to carry out his observations in Rio de Janeiro. The instruments were important agents in the establishment of communications between Sanches Dorta and Dawes, serving as intermediaries between both men.

Dawes reported that the Portuguese astronomers admired the quadrant, but that they were particularly impressed with the regularity of the clock. In this way, Dawes acknowledged the trained ear of the visitors in noting one of the central aspects of astronomical observations, namely the precise recording of the instant and the duration of celestial phenomena, which depended on the accuracy of the marking of time. This recognition of a shared observational skill is what assured Dawes's that a return visit to the Portuguese observation station in Rio de Janeiro was in his own interest.

Once there, Dawes copied the data collected by the Portuguese and would have copied more if time had allowed. Sanches Dorta, pursuing his own personal interests, was eager to exchange data with other savants, even if it meant sharing sensitive information such as his painstaking determinations of the longitude of Rio de Janeiro. By offering his observations, he took advantage of the opportunity to try to expand his range of relations, since he was interested in establishing correspondence with Nevil Maskelyne in Greenwich and as a result receiving data from the British observatory. In his narrative, Dawes himself is transformed into a type of intermediary agent between Sanches Dorta and the Astronomer Royal.

What is interesting in this episode is the possibility of reflecting on a type of network for the exchange of data collected all over the world that was formulated by and followed the movement of ships, operating across imperial borders. In this sense, it allows us to think about the inter-imperial connections in the construction of scientific knowledge carried out in colonial areas, especially in areas of intense maritime traffic.

Notes

1 Simon Schaffer, "Instruments, Surveys and Maritime Empire," in David Cannadine, ed., *Empire, the Sea and Global History: Britain's Maritime World, c.1763–c.1840* (Basingstoke: Palgrave Macmillan, 2007), 83–105, 86.
2 Letter from William Dawes to Nevil Maskelyne, Rio de Janeiro, September 3, 1787. Royal Greenwich Observatory Archives, Papers of the Board of Longitude, 14/48, 269–276.
3 See Jim Bennett, "The Travels and Trials of Mr. Harrison's Timekeeper," in Marie-Noëlle Bourguet, Christian Licoppe, and H. Otto Sibum, eds., *Instruments, Travels and Science: Itineraries of Precision from the Seventeenth to the Twentieth Century* (London: Routledge, 2003), 75–95; and several essays in Richard Dunn and Rebekah Higgitt, eds., *Navigational Enterprises in Europe and Its Empires, 1730–1850* (New York, NY: Palgrave Macmillan, 2016).
4 On Soares, see Heloisa Gesteira, "Instrumentos Matemáticos e a Construção do Território: a Missão de Diogo Soares e Domingos Capassi ao Brasil (1720–1750)," in Lorelai Kury and Heloisa Gesteira, eds., *Ensaios de História das Ciências no Brasil: das Luzes à nação independente* (Rio de Janeiro: EDUERJ, 2012), 207–224.
5 Jaime Cortesão, *Alexandre de Gusmão e o Tratado de Madrid* (Rio de Janeiro: Ministério das Relações Exteriores-Instituto Rio Branco, 1950-1960), pt. I, t. II, 12.
6 Maria Fernanda Bicalho, "A cidade do Rio de Janeiro e o sonho da Capital Americana: da visão de D. Luís da Cunha à sede do vice-reinado," *História* (São Paulo) 30, no. 1 (2011): 37–55; Junia Furtado, *Oráculos da Geografia Iluminista: D. Luís da Cunha e Jean Baptiste Bourguignon Danville* (Belo Horizonte: Editora UFMG, 2012).
7 Charles R. Boxer, *Salvador de Sá and the Struggle for Brazil and Angola* (London: The Athlone Press, 1952); Luís Felipe de Alencastro, *O Trato dos Viventes: Formação do Brasil no Atlântico Sul* (São Paulo: Companhia das Letras, 2000).
8 Bicalho, "A Cidade do Rio de Janeiro."
9 Richard Drayton, "Maritime Networks and the Making of Knowledge," in David Cannadine, ed., *Empire, the Sea and Global History. Britain's Maritime World, c.1763–c.1840* (Basingstoke: Palgrave Macmillan, 2007), 72–82, 72.
10 A detailed exposition of Sanches Dorta's life and career can be found in Ana María Marín Farrona, *Bento Sanches Dorta: representante portugués del progreso científico de la Ilustración* (Master's dissertation, University of Lisbon, 2011).
11 On the royal purchases of measuring instruments for Portuguese institutions at the time, see Rómulo de Carvalho, *A astronomia em Portugal no século XVIII* (Lisbon: Instituto de Cultura e Língua Portuguesa, Ministério da Educação, 1985); Isabel Malaquias,

"Instruments in transit: The Santo Ildefonso Treaty and the Brazilian Border Demarcation," in Marcus Granato and Marta Lourenço, eds., *Scientific Instruments in the History of Science: Studies in Transfer, Use and Preservation* (Rio de Janeiro: Museu de Astronomia e Ciências Afins, 2014), 99–113.

12 Jim A. Bennett, "The English Quadrant in Europe: Instruments and the Growth of Consensus in Practical Astronomy," *Journal of the History of Astronomy* 23 (1992): 1–14.

13 Letter from Martinho de Melo e Castro to D. Luis de Vasconcelos, Lisbon, February 17, 1781. Biblioteca Nacional do Rio de Janeiro, Manuscritos, MS 04, 04, 005.

14 Bento Sanches Dorta, "Observações Astronômicas Feitas Junto ao Castello da Cidade do Rio de Janeiro para determinar a latitude e Longitude desta Cidade," *Memórias da Academia Real das Sciencias de Lisboa* I (1797): 325–378, 325.

15 J. H. Magellan, *Description & Usages des instrumens d'Astronomie & de Physique, faits a Londres, pour Ordre de la Cour de Portugal en 1778 & 1779: adressée, dans une lettre, a son excélence M. Louis Pinto de Souza Coutinho* (London: Chez B. White, Libraire, en Fleet Street; P. Elmsley, Libraire, dans le Strand; & W. Brown, Libraire, au coin D'Essex Street, prés Temple Bar, 1779), 25.

16 Sanches Dorta, "Observações Astronômicas Feitas Junto ao Castello," 346.

17 For details on meteorological and magnetic observations see: A. M. M. Farrona, R. M. Trigo, M. C. Gallego, and J. M. Vaquero, "The Meteorological Observations of Bento Sanches Dorta, Rio de Janeiro, Brazil: 1781–1788," *Climatic Change* 115 (2012): 579–595; J. M. Vaquero and R. M. Trigo, "Results of the Rio de Janeiro Magnetic Observations 1781–1788," *Annales Geophysicae* 23 (2005): 1881–1887.

18 Sanches Dorta, "Observações Astronômicas Feitas Junto ao Castello," 325.

19 Ibid., 334.

20 Ibid., 326.

21 Bento Sanches Dorta, "Observações Astronômicas e Meteorológicas feitas na cidade do Rio de Janeiro no Anno de 1786 e 1787," *Memórias de Mathematica e Physica da Academia Real das Sciencias de Lisboa* III, Part I (1812), 68–153, 108.

22 Bennett, "Travels and Trials."

23 One such astronomer was the French Nicolas Louis de la Caille, who had been to Rio de Janeiro in 1751, during a trip to the Cape of Good Hope. See his *Journal Historique du Voyage fait au cap de Bonne-Esperance* (Paris: Chez Guillym, 1763).

24 David Philip Miller has demonstrated how the business of longitude measurement depended crucially on formal and informal networks for exchanging data between places spread out over vast regions. See David Philip Miller, "Longitude Networks on Land and Sea: The East India Company and Longitude Measurement 'in the Wild', 1770–1840," in Richard Dunn and Rebekah Higgitt, eds., *Navigational Enterprises in Europe and its Empires, 1730–1850* (New York, NY: Palgrave Macmillan, 2016), 223–247.

25 Letter from William Dawes to Nevil Maskelyne, f. 269v.

26 Ibid., f. 273r. Although Dawes notes the presence of two astronomers, he refers only to Sanches Dorta by name, at the end of the letter.

27 Schaffer, "Instruments, Surveys and Maritime Empire," 93.

28 Letter from William Dawes to Nevil Maskelyne, f. 273v.

29 Ibid.

9 Auguste de Saint-Hilaire's Writings between European and Brazilian Audiences, 1816–1850

Lorelai Kury

Botanical Rivalries

The travel account and botanical writings of the French naturalist Auguste de Saint-Hilaire (1779–1853) are a most valuable source of information for Brazil's natural and social sciences. These textual artifacts are the product of complex collective processes that engaged a network of people, plants, letters, journals, newspapers, books, ideas, opinions, and controversies connecting the Brazilian and French empires. Brazilians were an object of study for Saint-Hilaire, but they were also agents who proved essential to his travels and the production and reception of his work.

Saint-Hilaire traveled to Brazil in 1816 in the company of the Duke of Luxembourg, who was on a diplomatic mission to the court of king D. João VI in Rio de Janeiro, who had fled the Napoleonic invasion of Portugal years before and took residence in the capital of his South American colonial possession. The duke welcomed Saint-Hilaire along because the two men were connected by family ties. The French naturalist also obtained funding from the Ministry of the Interior, who consulted with the Muséum d'histoire naturelle in Paris about the candidate's abilities.

During his stay in Brazil, which lasted until 1822 (the year when D. João VI's son declared independence from Portugal and styled himself the emperor of Brazil), Saint-Hilaire regularly sent natural products back to France, chiefly animal and plant specimens. Most notably, he prepared dried plants to form an herbarium. Although the dried specimens were first shipped to the natural history museum, from there they were forwarded to the traveler's family. Thus Saint-Hilaire ensured his ability to capitalize on the botanical findings of his trip since only he would have access to the specimens, as well as to the preliminary descriptions drafted in the field. Other types of natural products stayed in the museum.

Everything published by Saint-Hilaire in the years immediately following his return to France drew from the material he had compiled in Brazil. He was quick to publish excerpts from his travel diary and the botanical treatise *Histoire des plantes les plus remarquables du Brésil et du Paraguay* (1824). In order to guarantee primacy in the description and classification of genera and species, his strategy was to start from the plant groups he knew best. Competition had grown quite fierce. Naturalists in the German-speaking world had already begun publishing about Brazilian plants, and in the scientific milieu of the time, primacy in the description of

DOI: 10.4324/9780367814540-11

species and groups could open important doors to jobs and institutions, as well as promote the author's renown and authority.[1] There was a worldwide competition to publish on Brazilian flora, involving dozens of naturalists and collectors who had traveled across Brazil, especially after 1815. The early 1820s saw a flurry of works on Brazilian natural history, based on expeditions undertaken in the previous years. Saint-Hilaire told his collaborators that "the Germans" had "taken" some species from him.[2] He advised them to "print in some periodical" the descriptions of the plants with which they were working so they would not have to change their names.[3]

According to Swiss naturalist Augustin-Pyramus de Candolle, writing in the late 1820s, a general estimate of European herbaria indicated that the number of known plant species had climbed from 500 at the dawn of the nineteenth century to 14,000. He warned, however, about a danger to be avoided: "the numerous publications being released in four or five different cities simultaneously should not overlap, and the same objects should not be described here and there by different names."[4] With this in mind, he exhorted the botanists who were leading these studies to communicate often, try to avoid duplicate work, and respect names already given. This would be no problem, as he saw it, given the supposed spirit of fair play and loyalty displayed by the naturalists.

Indeed, the internationalization of science had become unstoppable. The circulation of journals was a key feature of this trend, alongside epistolary exchanges. Naturalists circulated as well, on scientific visits and conducting research at government-sponsored and private collections. And within the "Republic of Science," these exchanges occasioned fiercer competition in the quest for scientific recognition.[5] But Candolle adverted that an individual's personality bears as much on the achievement of success as does talent. Therefore, passing judgment on the work of fellow naturalists William John Burchell (1781–1863), René Louiche Desfontaines (1750–1833), and José Correa da Serra (1750–1823), Candolle identified their respective traits of indecisiveness, hesitancy and insouciance as factors that kept each one from realizing their full potential. About Saint-Hilaire, Candolle wrote that his "finicky nature contributed to diminishing what one could expect of his work."[6]

Indeed, Saint-Hilaire's over-attention to detail rests at the heart of many of his academic clashes, including his uneasy collaboration with Adrien de Jussieu and Jacques Cambessèdes in the book *Flora Brasiliae Meridionalis*, published between 1824 and 1833. He criticized the two young naturalists for not knowing words in Portuguese and locations in Brazil. The two collaborators also had to deal with Saint-Hilaire's possessiveness over his herbarium and notes.[7] Although he himself entrusted this work to people who had never set foot in Brazil, he acted as if his experience and memories of events meant he was the only one capable of undertaking the work of describing and classifying Brazilian plants. At one point, he required the two assistants to return his notes since, as he put it, they apparently thought it was enough to have a "microscope and good eyes" to see what he had seen in person.[8]

Saint-Hilaire's knowledge of Portuguese, detailed notes, rigorous method, and contacts with Brazilians, therefore, became like witnesses summoned to testify on his behalf. Yet it was his thoroughness that fed into his strategy for claiming authority about Brazilian topics. In the foreword to his *Voyage aux sources du rio de*

S. Francisco et dans la province de Goyaz, published in 1848, he said that Brazilians had encouraged him to publish the volume and that they were better judges than Europeans when it came to matters of their concern.[9] He justified his methods in the same foreword and noticed that he had exercised great rigor in referencing his sources and citing authors. He also remarked that a Brazilian gentleman writing in the periodical *Minerva Brasiliense* had said it would be of greater benefit to correct and cement known geographic and ethnographic facts about Brazil than to go after new ones. His obsession about getting names right in Portuguese may have had something to do with his desire to be known as the most reliable and correct among European travelers writing about Brazil. In the same foreword, he went on to write that he had decided to point out mistakes by other authors who had commented on the same places that he had described because he wished to be useful to Brazilians.

Saint-Hilaire made a target out of himself when he reproached and corrected his rivals in his travel accounts. Prince Maximilian of Wied-Neuwied (1782–1867) felt so attacked by his criticisms that in 1850 he published an addendum to his travel narrative, both to update information on Brazil and to respond to Saint-Hilaire. The prince, a naturalist and ornithologist, had traveled across Brazil from 1815 to 1817, traversing stretches between Rio de Janeiro and Bahia. His observations about the Botocudo Indigenous people and his work on zoology were well received. His travel account was published in German in 1820 (*Reise nach Brasilien*) and was soon translated into both English and French. It was only released in Portuguese in 1942.

In his addenda, Wied-Neuwied listed Saint-Hilaire's references to his work and accused the French botanist of singling him out as his favorite target. Saint-Hilaire criticized such things as Wied-Neuwied's localization of towns and rivers, descriptions of people's customs, knowledge of toponyms, and names of plant species. In one footnote, for example, Saint-Hilaire wrote: "I must admit it is hard for me to recognize the stretch that I traveled between São João and Macaé as described by the Prince of Neuwied."[10]

Examples like this are abundant. The prince responded to a number of these criticisms, generally by citing local informants as his sources or casting blame on the poor French translation of his book. In some cases – speaking about cattle-raising in the Brazilian province of Minas Gerais, for instance – Wied-Neuwied agreed that his colleague might be right, because he had not confirmed the data himself but had relied on someone else's information.

Part of Saint-Hilaire's qualms with Wied-Neuwied had to do with Portuguese words and place names. It should be "Bragança" and not "Bargaņca," "Saquarema" instead of "Sagoarema," "Ceri" and not "Ciri," and so on. The prince was irritated by this type of nitpicking and defended himself by saying that his ears had heard it that way[11] or, in one case, by writing: "Mr. A. de Saint-Hilaire, […] it would appear, considers himself the only one who heard these Brazilian names and learned to pronounce them well."[12]

This minor controversy within the Republic of Science not only evinces Saint-Hilaire's penchant for hair-splitting and his somewhat unforgiving nature; it also shows how the information contained in a travel account could be jealously controlled. In addition to the fieldwork undertaken during trips, much work was put into assembling publications on Brazil, either about taxonomy or travel narratives.

When composing their reports, travelers relied on notes taken in the field but also turned to available publications and to Brazilian interlocutors to whom they had ongoing access. Wied-Neuwied criticized Saint-Hilaire's transcriptions from the language of the Botocudos. The prince's authority here was unassailable, for he could rely on the knowledge of Guäck (Kuêk), an Indigenous man whom he had taken back to Neuwied and to whom he referred as "his" Botocudo.[13] Firmiano, the young Botocudo whom Saint-Hilaire wanted to "adopt," ended up staying in Brazil after the two men had a number of run-ins.[14] So at least in this respect, the prince came out on top.

In European scientific circles, it had therefore become challenging to be recognized as an expert on Brazilian nature and descriptions of Brazil. Saint-Hilaire felt especially threatened by the German-speaking naturalists who had visited Brazil around the same time as him, Carl Friedrich Philipp von Martius (1794–1868) above all. In addition to remonstrating Martius on matters of botany, Saint-Hilaire was critical of the collective nature of his whole enterprise (the *Flora brasiliensis*, which Martius started to publish in 1840, to be completed only several decades after his death), the grandiosity of his goals, and his haste – and thus superficiality – when it came to publishing. As to Wied-Neuwied, Saint-Hilaire homed in on mistakes in place names. As he explained, precise transcription of the words used by the locals were necessary to ensure geographical certainty. Moreover, he longed to be the best and most precise travel narrator. In this regard, addressing a friend in 1847, Saint-Hilaire boasted of the reputation he had built among Brazilians:

All of these Brazilian books carried me back to a happy past; I once again saw my lovely blue sky, my lovely deserts, the places I had watered with my perspiration; through these books I received news of my friends and, I will tell you, I was greatly pleased to see how, because of [my books'] accuracy, they place me at the helm of the travelers who traversed Brazil.[15]

Today, Saint-Hilaire and Martius stand among the most prominent of the nineteenth-century naturalists who journeyed across Brazil, because of both their botanical work and their travel accounts. While the nineteenth-century Brazilian elite was aware of the account written by Martius and his companion Johann Baptist Spix (1781–1826), thanks to the former's personal ties with Brazilians and his many contributions to learned periodicals published in the country, the book was only translated into Portuguese in full in 1938. It is likely that a linguistic barrier worked in Saint-Hilaire's favor, since Brazil's nineteenth-century elite was as a rule educated in French, but few of its members were familiar with English and even fewer with German.

Sources and Informants

We can see that rivalries among European botanists who studied Brazilian flora had repercussions overseas. When it came to Brazilian topics, information from the Brazilian press and other "opinion makers" based in the country carried as much weight as that coming from members of European scientific and literary circles. A vital source of credibility for Saint-Hilaire was ostensibly displaying that he was permanently following the debates carried out at the Brazilian Historical and Geographical Institute, founded in 1838, pronouncements by the Brazilian emperor and members of the country's political elites, and information coming from the periodical press and printed works by Brazilians.[16] Reciprocally, he never failed to send copies of his books to influential individuals in Brazil.[17] Now, it is true that until the twentieth century, Brazilians were seldom involved in the botanical debates that increasingly took place in specialized scientific journals. European botanists could master what little information was available in printed sources relatively quickly. In a letter to his collaborators Jussieu and Cambessèdes, Saint-Hilaire listed the sources on Brazilian plants that were available at the time, starting with the seventeenth-century work of the Dutch naturalists Wilhelm Piso (1611–1678) and Georg Marcgraf (1610–1644). To this he added the works of Domenico Vandelli (1735–1816), director of the botanical gardens of Ajuda, in Lisbon; those of Brazilian naturalist Manuel Arruda da Câmara (1752–1810) and the Portuguese Bernardino Antônio Gomes (1768–1823); and some scattered sources of plant descriptions. He finished offering an overview of what European naturalists – his competitors and rivals – had recently published.[18]

More general information on Brazil, about its geography, history, and customs, could be sourced from a larger number of published works. Saint-Hilaire consulted works from such authors as the Portuguese geographer Manuel Aires de Casal (1754–1821), who lived in Brazil for a long time, or the Brazilian historian José de Sousa Azevedo Pizarro e Araújo (1753–1830), along with short-lived Brazilian newspapers like *O Patriota* (1813–14) and *Minerva Brasiliense* (1843–45). Gaining access to these sources was precisely one of his challenges. In the introduction to his volume on his travels to the headwaters of the São Francisco River and to the province of Goiás, he explicitly acknowledges the assistance that some Brazilians and Frenchmen had given him in obtaining works published in Rio de Janeiro, Pernambuco, and São Paulo. They included José de Araújo Ribeiro, a diplomat in Paris; Pedro de Alcântara Lisboa, a Paris-trained Brazilian chemist who was well connected to the French embassy; Joseph-François-Xavier Sigaud, a French physician at the Brazilian imperial court; and Ferdinand Denis, conservator of the Sainte-Geneviève Library in Paris.[19]

In a catalog of the scientific works owned by Saint-Hilaire that were donated to the city of Montpellier, several items that he quotes in his own publications are missing from the list, including the works cited earlier. We do not know if they were donated to the Solesmes Abbey, kept by family or friends, or lost or stolen. The list shows that Saint-Hilaire's library held some books by Brazilian and Portuguese writers, acquired over the course of his lifetime. The catalog features booklets by Domenico Vandelli and Arruda da Câmara, along with mid-nineteenth century works by authors like the Viscount of São Leopoldo; Fernando Antônio Pereira de Vasconcelos, director of the Ouro Preto botanical garden; and Brazilian botanist Francisco Freire Allemão. Further listed are such general reference works as *Annuario politico, historico e estatistico do Brazil* (1846–1847), collections of statistics, and speeches given by governors of the province of São Paulo in the 1840s. Saint-Hilaire also owned an 1827 edition of José Mariano da Conceição Veloso's *Flora Fluminensis*.[20]

The catalog mentions a number of medical theses defended at the Paris and Montpellier schools of medicine by Brazilians such as José Martins da Cruz Jobim, Manuel de Mello Franco, and Tomás Gomes dos Santos. Brazilians studying in France probably constituted one of the most common sources of news on Brazil, an example being Gomes dos Santos, cited in Saint-Hilaire's personal correspondence with botanist Alfred Moquin-Tandon.[21] We know that Saint-Hilaire wrote letters of recommendation for Piotr Czerniewicz (Pedro Chernoviz), a Polish physician who graduated from Montpellier in 1837 and who went on to write a classic Brazilian medical manual in the latter half of the nineteenth century.[22] Saint-Hilaire's correspondence notes the presence of various members of the Brazilian elite in France, especially in Paris.

Reciprocally, many Brazilians visiting France carried letters of recommendation addressed to Saint-Hilaire, like a young "mulatto" traveling in the company of his tutor on his way to study singing in Italy, who, according to a sarcastic remark by Moquin-Tandon, sang like a frog and had a "*je ne sais quoi* of Botocudo about him."[23] Diplomats and a few Brazilians who were just passing through also ended up meeting Saint-Hilaire, who was considered a kind of "friend" of Brazil – so much so that the Brazilian emperor made him a knight of the orders of Christ and of the Southern Cross.

In one of his studies, the French hygienist Alexandre Parent-Duchâtelet, Saint-Hilaire's brother-in-law, stated that he received two important figures from Brazil, Francisco Jê Acaiaba de Montezuma and Manoel Theodoro de Araújo Azambuja, who were both interested in getting to know French public health institutions, with a view to developing similar initiatives in Brazil.[24] One of Azambuja's biographers wrote that Saint-Hilaire helped Azambuja settle with his family temporarily in Paris in 1824. Saint-Hilaire did this out of gratitude for the fine reception Azambuja had accorded him during his own Brazilian stay.[25]

Details of these personal encounters did not always make it into the records. A somewhat more reliable way to chart Saint-Hilaire's use of information coming from Brazilian and Portuguese sources is by examining the catalog mentioned earlier and the references found in his travel accounts, scientific works, and correspondence. The forewords and footnotes in his writings provide important evidence of the texts he used, as well as of his method, which consisted primarily of comparing his own annotations with those of as many other trustworthy authors as possible. He had a clear preference for some writers whose accuracy he valued, like Aires de Casal and certain European travelers, such as the Scotsman George Gardner.

Saint-Hilaire's diligence in citing and referencing Brazilians was part of his effort to enhance his relations with the Brazilian elites, while it was also a way of showing how up-to-date he was. Since he spent long seasons in Montpellier or at his property near Orléans, he relied on friends or disciples in Paris to gather news about Brazil. His disciple Charles Naudin had Brazilian books shipped to Paris via Lisbon. But some major sources often lay outside his grasp. In a letter addressed to Ferdinand Denis in 1847, Saint-Hilaire justified his lack of familiarity with recent publications:

> You are right; the distance that my ill health has forced upon me has particularly intensified the bibliographic difficulties that I would perhaps have encountered even in Paris. Montpellier is considered the second scientific city in the realm, but I would challenge anyone, while living in it, to write even two lines about the vegetation of Asia or that of South America.[26]

In some cases, we see how swiftly Saint-Hilaire employed fresh data. After telling Naudin that he was not adding any more footnotes to his book *Voyage aux sources du rio de S. Francisco et dans la province de Goyaz*, and sending the manuscript off to his editor in Paris, Saint-Hilaire found time to insert a few more things – for instance, references to *Itinerário do Rio de Janeiro ao Pará*, published in 1836 by Raimundo da Cunha Matos, a Brazilian military officer and founder of the Historical and Geographical Institute. In 1847, Denis lent Matos's book to the Saint-Hilaire, along with some other volumes.[27] It was imperative for Saint-Hilaire to demonstrate his familiarity with recent Brazilian works in order to safeguard the name he had forged for himself in the disputed literary space of travel accounts. And in more than one occasion, he underscored the fact that Brazilian readers were important to the French publishing market. He wrote to Naudin, for example:

> I just received a letter from the physician of the Emperor of Brazil, where he tells me, "You are cited by Brazilians as the most truthful, most well-informed traveler about places." So, it is likely that some copies of my Goyaz will be sold in Brazil.[28]

Differentiating and Ranking

In his correspondence and publications, it is apparent that Saint-Hilaire took great pains in describing his relations with powerful Brazilian landowners and merchants, and also with members of the government and of the intellectual, civil, military, and religious elites. As to the general public, while he thought their customs and behavior were not very "civilized," he did recognize positive qualities in certain individuals and provinces, mentioning, for instance, the hospitality of the residents of Minas Gerais. To a certain extent, he used his travel accounts to thank people for their warm welcomes and assistance. For example, he wrote a few lines of gratitude to a parish priest in Santa Luzia, Goiás, who had proven knowledgeable and attentive and who had likewise earned praise from another traveler (Johann Emanuel Pohl): "Should these pages find their way to Mr. João Teixeira Alvarez, he will see that two foreigners whom he welcomed under his roof have fond memories of his virtues, and that honest men of all nations, united in sweet brotherhood, are able to understand, appreciate, and like each other."[29]

While Saint-Hilaire's tone is generally quite critical of Brazilians, there are abundant exceptions to this rule. On occasion, he has words of praise for people he met along the way who displayed commendable qualities; thus, he is able to commend individuals from groups that he otherwise felt to be characterized by laziness, irrationality, a lack of altruism, and immorality. To cite one such case, near the rural Minas Gerais village of Capelinha, he was warmly received by "mulattoes who appeared to be quite poor." They offered Saint-Hilaire and his traveling companions the best of what they had and charged nothing for it.[30]

Alongside Saint-Hilaire's relations with the general public, his status as a man of science opened doors to connections of trust and exchange among peers. While the French naturalist evidently felt that his nationality and the fact that he frequented Parisian institutions of natural history imbued him with superiority, he recognized people who had studied in Europe as belonging to the same category his, such as Friar Leandro do Sacramento, José Bonifácio de Andrada e Silva, and Manuel Ferreira da Câmara de Bittencourt e Sá.

In his personal relations with Portuguese and Brazilians, he also adopted another important strategy, which was to draw a clear distinction between social groups and, within the ruling strata, to differentiate the categories of people with whom he had dealings. As an aristocrat, he could approach and gain the protection of noblemen from the top echelons of the Portuguese government, such as the Count of Barca (Antônio de Araújo e Azevedo). He also enjoyed decisive travel assistance from military officers, who played a key role in the administration of the immense Brazilian territory.

Portuguese gentlemen and government agents would at times "concede favors" and could also be "generous." According to Saint-Hilaire, he strived to be always courteous toward them. In December 1816, he wrote in a letter to his friend Deleuze:

> With a promptness and goodwill that I cannot praise enough, the Portuguese government granted me all the passports that I needed for my travels through Minas. Not long before, I had obtained permission to stay at the King's palace in Santa Cruz and take everything that would be necessary for my sustenance; and you can imagine, *Monsieur*, that I did not ask for this. Foreigners from another nation have shown ingratitude for the benevolence displayed by the Portuguese, who have always treated them kindly; I shall endeavor not to follow such an example. If there is anything that could make up for the distance from all those whom we hold dear, it is certainly the demonstration of kindness and esteem that I have received from the Portuguese and the Frenchmen whose business brings them here every day.[31]

The Portuguese and Brazilians often complained about the inaccurate ideas that foreigners had about Brazil. For example, Brazil's first official scientific expedition throughout its own territory, in 1859, was meant to assess the empire from a Brazilian perspective and refute the views of foreigners who had only superficial knowledge of the country.[32] The goal was to arrive at a more balanced view of Brazil, underscoring its civilized characteristics and downplaying controversial matters, such as slavery and the population's lack of "civilized customs."

By the early nineteenth century, at the time of Saint-Hilaire's travels, Brazilian elites had already acquired a sense of nativist pride. In 1813, the editor of *O Patriota*, a newspaper published in Rio de Janeiro, criticized the British writer Andrew Grant, who had penned a very caustic portrait of Portuguese-Brazilians. In his book, *History of Brazil, comprising a geographical account of that country* (1809), Grant laid stress on the immorality of the inhabitants of Rio de Janeiro. The editor of *O Patriota* replied that Grant had mistaken "rabble" for decent people. According to the paper, the people Grant saw on the streets were not true Portuguese-Brazilians, but persons of African descent, from the so-called "barbaric nations."[33]

Indeed, this was the core of the identity question for the Portuguese-Brazilian elite at that time: that they should not be confused with enslaved Africans, free poors or Indigenous peoples. Even great landowners considered slavery a "barbaric" custom, but the arguments used to justify it were the need for manpower and the fact that, according to advocates of slavery, Africans were treated much better in Brazil than working-class white persons were in Europe.[34]

Saint-Hilaire did not conflate these groups. His concern with distinguishing groups among Brazilians is apparent in his correspondence with Cambessèdes during the correction of the manuscript of *Plantes Usuelles des Brasiliens*, published in 1827. Saint-Hilaire complained that young naturalists had used the term "Brazilian" to refer to Indigenous peoples. Cambessèdes's reply suggests that this type of matter did not worry him much:

These days, Brazilians do not see themselves as the primitive inhabitants of their country and it would not be fair for them to deny this name to the first inhabitants of the land in which they are now living. I cannot see how it would be possible to confuse one with the other.[35]

In addition to differentiating between wellborn Portuguese-Brazilians and those who were not, Saint-Hilaire also learned to display his own social standing. According to his account, on one occasion when traveling through the northern region of Rio de Janeiro, about the village of Campos dos Goytacazes, he visited a former Jesuit college and was invited to the table but treated in a cold manner. He later realized this was because he had failed to groom and dress appropriately. He was wearing plain, light clothing, which he considered best suited to long travels in a hot climate. As he saw it, in Brazil more than anywhere else, "men of high class assign great value to attire." He devised a simple strategy for fitting in under the various circumstances that he might encounter:

As I was familiar with the country's habits, while not wishing to deprive myself of the advantages that light, cheap vestments afford a traveling naturalist, I took care to place appropriate clothing in the top part of a suitcase and, before presenting myself at the homes of the well-to-do, to complete my grooming in the shade of some tree.[36]

Strict control of etiquette and formal protocol may indeed be crucial during cross-cultural interactions, as Sanjay Subrahmanyam pointed out in another context.[37]

On several occasions, Saint-Hilaire was not accorded the treatment he wanted. In these situations, his friendship ties, letters of recommendation or passport ensured him at least some measure of comfort. For instance, doors were opened thanks to his friendship with João Rodrigues Pereira de Almeida, a prominent merchant and slave trader. According to Saint-Hilaire, Almeida's name was so powerful that it was like a lucky charm: "This name, which has so often been a talisman for me, once again had its usual effect. They soon treated me with great respect, offering me breakfast, with coffee, milk, bread, and butter."[38] His French passport exempted him from paying tolls at bridges or other government barriers. He once asked to lodge at the home of an authority in a small village in the province of Espírito Santo but was turned away. After a brief conversation during which Saint-Hilaire presented his passport, his interlocutor changed his behavior quite dramatically: "The respect that Brazilians have for their superiors was such that a mere glance at the signature of the minister of state Thomaz Antônio de Villanova e Portugal had the effect of a magic phrase. The home became mine, everyone was at my service, and they insisted that I stay."[39] Letters of recommendation were also highly useful during the traveler's journey; these were addressed to specific people living in villages and more prosperous settlements as well as in rather remote regions. To cite one of the myriad examples: "I had a letter of recommendation for the owner of S. Elói, Captain Pedro Vergiani; no one welcomed me more cordially, and I spent a day at his home."[40]

So, despite his somewhat difficult personality, Saint-Hilaire learned to circulate among Brazilians and be welcome. Above all, he learned how to set people apart and be set apart. However, he also needed to negotiate local patronage networks[41] and remember that his travel routes, lodgings, shipment of plants to France, access to credit, and other travel needs were dependent on the relationships of trust and camaraderie he managed to establish – not only with powerful individuals but also in his direct contact with simple country folk and his immediate aides.[42] In a letter to the Muséum dated 1819 (that is, while he was in Brazil), he said he had discovered while there that the best way to deal with Brazilians was to establish friendly ties with them.[43]

Some of the "friendly ties" he forged with Brazilians over the course of his travels were long-lasting and impacted his written production. Almost 30 years after returning to France, he confessed to his disciple Charles Naudin that he had moderated his comments on the degeneration suffered by white Europeans in Brazil: "In writing about Brazil, I tempered this failing as best I could." According to him, white immigration would not be an efficient way of "revitalizing those people," since "the Caucasian race tends to deteriorate and the African race, to perfect itself" in the part of America he had visited. However, Saint-Hilaire admitted that there were "attenuating circumstances" for this deterioration, which he had enumerated in his book about his travels in São Paulo.[44] He was likely referring to the climate in the province of São Paulo, which he deemed more like Europe's and unfavorable to the health of Africans.[45] We know nothing of the content of the letter initially sent to Naudin, but Saint-Hilaire's responses indicate that, in this particular case, his negative appreciation referred primarily to Brazil's climate.[46] However, in a number of his writings, we can observe that he also considered Brazilians to be inferior to Europeans because of their constituent "races." About São Paulo, for example, he remarked sharply that many residents who at first looked white in fact had Indigenous blood as well and displayed the traits of these forebearers, especially "negligence."[47]

Publishing on Brazil

A letter dated April 1828 reveals that Saint-Hilaire was asking Pierre-Jean-François Turpin to help him find an editor for what he called the "historical narrative" of his travels in Brazil.[48] At the time, Saint-Hilaire was in Montpellier under the care of his friend Michel-Félix Dunal, physician and botanist. Turpin, who started his career as a military officer, turned to natural history and botanical illustration, traveling across Central America and the United States. Back in Paris, he published scientific memoirs based on Goethe's theories of botany and also illustrated dozens of natural history books, including part of *Flora Brasiliae Meridionalis* (1824–1833). He was elected to the Académie des Sciences in 1833.

Turpin reported first contacting Panckoucke, the publishing house that had released the *Encyclopédie méthodique* and was at the time finishing the second edition of the colossal *Description de l'Égypte* (1821–1830), which contained 11 illustrated atlases. In collaboration with Jean-Louis-Marie Poiret, Turpin had published the lavishly illustrated *Leçons de Flore* with Panckoucke. Turpin provided a disparaging, sarcastic description of his talk with the publisher, who showed no interest in publishing Saint-Hilaire's work. According to Turpin, professionals in Paris had to decide on a single branch of work, and this was true of booksellers and editors as well, who had to choose one type of audience and develop an "atmosphere" for their business. "Panckoucke's atmosphere-clientele consists of works that never comprise fewer than sixty volumes." Furthermore, Turpin went on, the publisher's catalog showed that their authors were poorly skilled, easily fired, and easily replaced. A lot of money was spent on flyers, newspaper advertisements, and dinners; when the public had been drawn in, the publisher turned on the machine and made a lot of money: "It's a big machine, one that no longer even busies him and that produces gold, gold, and more gold."[49] In other words, even though Saint-Hilaire's work enjoyed ministerial support, it was of no interest to the publisher, whose eyes were focused mainly on the market for large illustrated books, with in-folio atlases.[50]

Turpin next reached out to the Levrault publishing house, represented in Paris by Pitois. The conversation was more encouraging this time, as the booksellers inquired about monetary expectations and whether or not the book would be illustrated, and asked for a few days to think it over. But Pitois made it clear from the outset that he thought a multi-volume historical narrative was too long. Turpin wrote to Saint-Hilaire suggesting that he get in touch with the bookseller Arthus Bertrand, whose "business atmosphere" was limited to travel narratives, just in case negotiations with Levrault fell through. There are no details on later negotiations. The first volumes of Saint-Hilaire's work were published by Grimbert et Dorez and by Gide. Arthus Bertrand released the later volumes.

Publishers at the time wanted to cut on text and increase the number of images.[51] Saint-Hilaire's travel accounts feature only two illustrations: one showing the house where the Duke of Luxembourg stayed in Rio de Janeiro and one showing one of Saint-Hilaire's aides, the Botocudo Firmiano. This scant number of images is probably due to the fact that Saint-Hilaire was a bad draftsman. Even his strictly scientific work was illustrated by others, something his critics and rivals liked to point out. When he was elected to the Académie des Sciences, for example, Charles-François Brisseau de Mirbel labeled his lack of artistic skill a flaw.[52] Some naturalists who traveled across Brazil around the same time as Saint-Hilaire published beautifully illustrated picturesque historical narratives, such as Spix and Martius and Prince Wied-Neuwied.[53] This iconographic material still stands as a valuable reference in representations of landscapes, the presence of Indigenous peoples, and customs in what was then a Portuguese colony. Despite the dearth of images in Saint-Hilaire's historical narrative, his botanical work was illustrated by excellent artists, mainly Turpin, albeit very few editions are colored. We know that Brazilians had subscriptions to his works published in France and there was at least one subscriber to the lavish hand-colored edition of *Flora Brasiliae Meridionalis* from Rio de Janeiro.[54]

Without the appeal of images, Saint-Hilaire's historical narrative of his trip relied on the verbal richness of his textual descriptions. His prose is very fluid in the volumes published in the 1830s and offers a number of poetic or picturesque passages. Later volumes have a greater amount of dry, statistical-like content, with numerous lists and tables and lengthy descriptions of the administrative structures in the places he visited.[55] His status in the travel literature market derived from his commitment to reliability and the compilation of information, which he drew from his own notes and the sources he managed to consult. And because of his status in the French scientific and cultural milieu and his skill in dealing with the Brazilian imperial elites, his vision of Brazil became a kind of civilized reference point for evaluating the features of the former Portuguese colony.

What we can glean from the scant information available on Saint-Hilaire's negotiations with publishers is that the first and second parts of his account, published in 1830 and 1833, sold well, yielding him some profit: 1,000 francs and 500 francs, respectively.[56] Despite government subsidies (i.e., the guaranteed purchase of one hundred copies), publishers were much more reluctant to take on the third part. As Saint-Hilaire saw it, the French public had had its fill of travel books:

> I count on Brazil more than France; in France, people are tired of travels, of the railroad almanac that costs 25 c; the *Guide des voyageurs* is for the erudite; yet I feel that even in Brazil, where French newspapers arrive, the book must be announced in French newspapers.[57]

As the publication of relatively inexpensive books grew, Saint-Hilaire's work became affordable by only a few. The Brazilian public, he argued, had a special interest in books about their own country, and they likewise looked for the latest things from France, which extended to their reading the French periodicals that arrived at Brazil's main ports.

Conclusion

Saint-Hilaire thus produced his scientific and literary work in a context of rivalry with other botanists and in a competitive market for travel accounts. I believe that the Brazilian audience was a key source of legitimacy for the Europeans who wanted to become experts on the country's nature or customs. Emperor D. Pedro II was undoubtedly an essential figure here, bestowing honorary titles, acquiring books, issuing public statements, and, in some cases, supporting publications financially.[58] In addition to the emperor, Saint-Hilaire's potential reading public included the lettered elites – landowners, large merchants, liberal professionals, military officers, and top-ranking clerics – all of whom could lend legitimacy to his narrative.

Saint-Hilaire's writings always evince a concern with demonstrating that he was well-versed in the language and customs of Brazil. He also expressed his gratitude to those who received and welcomed him so warmly, going so far as to soften his opinions on the Brazilian climate and the prevailing mixing of the races. After returning to France, he counted on Brazilians as readers as well. Brazilians were therefore not just the object of his research; they were also informants, judges of his accuracy, and readers of his work.

Notes

1 On the construction of scientific authority and credibility, see Steven Shapin, "Trust, Honesty, and the Authority of Science," in Ruth Ellen Bulger, Elizabeth Meyer Bobby, and Harvey V. Fineberg, eds., *Society's Choices: Social and Ethical Decision Making in Biomedicine* (Washington, DC: National Academy Press, 1995): 388–408; and Shapin, *Never Pure: Historical Studies of Science as if It Was Produced by People With Bodies, Situated in Time, Space, Culture, and Society, and Struggling for Credibility and Authority* (Baltimore, MD: Johns Hopkins University Press, 2010).

2 Letter, Saint-Hilaire to Adrien de Jussieu, Orléans, June 5, 1827. Muséum national d'histoire naturelle (MNHN), ms. 53.

3 Letter, Saint-Hilaire to Adrien de Jussieu, Orléans, March 5, 1827. MNHN, ms. 53.

4 Augustin-Pyramus de Candolle, "Notice sur la botanique du Brésil," *Bibliothèque Universelle* (1827): 1–11, here 10–11.

5 For a discussion of the notion of "Republic of Science," see "La République des sciences," special issue of *Dix-Huitième Siècle* 40, no. 1 (2008).

6 Augustin-Pyramus de Candolle, *Mémoires et souvenirs de Augustin-Pyramus de Candolle. publiés par Alphonse de Candolle* (Geneva: J. Cherbuliez, 1862), 414.

7 Lorelai Kury, "Auguste de Saint-Hilaire: la botanique et l'expérience du voyage," in Claudia Damasceno Fonseca, Laura de Mello e Souza, Michel Riaudel, and Antonella Romano, eds., *Le moment 1816 des sciences et des arts. Auguste de Saint-Hilaire, Ferdinand Denis et le Brésil* (Paris: Sorbonne Université Presses, 2022): 221–244.

8 Letter, Saint-Hilaire to Adrien de Jussieu, n.p., [January 7, 1830]. MNHN, ms 53.

9 Auguste de Saint-Hilaire, *Voyage aux sources du rio de S. Francisco et dans la province de Goyaz*, 2 vols. (Paris: Arthus Bertrand, 1847), vol. 1, foreword, vi. The book is dated 1847, but the foreword is from 1848.

10 Maximilian Prinz zu Wied, *Brasilien. Nachträge, Berichtigungen und Zusätze zu der Beschreibung meiner Reise im östlichen Brasilien* (Frankfurt: Heinrich Ludwig Brönner, 1850). I have used the Brazilian edition: Príncipe Maximiliano de Wied, *Acréscimos, correções e notas à descrição de minha Viagem pelo Leste do Brasil* (Rio de Janeiro: Conselho Nacional de Pesquisa, 1969): 82.

11 Wied, *Acréscimos,* 32.

12 Ibid., 18.

13 On the repatriation of the skull of Kuêk, see Ludimila de Miranda Rodrigues, *Paisagens culturais alternativas no Brasil contemporâneo e vivência espacial da comunidade indígena Krenak* (Master's thesis, Universidade Federal de Minas Gerais, 2013): 151–153. On Maximilan, see Renate Löschner, ed., *Viagem ao Brasil do Príncipe Maximiliano Wied-Neuwied: Biblioteca brasiliana da Robert Bosch Gmbh* (Petrópolis: Kapa Editorial, 2001).

14 Auguste de Saint-Hilaire, *Voyage à Rio-Grande do Sul (Brésil)* (Orléans: H. Herluison, 1887), 595.

15 Letter, Saint-Hilaire to Charles Naudin, Montpellier, February 11, 1847. Wellcome Library, ms. 7572.

16 On Brazilian imperial institutions and the role of emperor D. Pedro II as a patron of artists and scientists, see, among others: Georges Raeders, *Pedro II e os Sábios Franceses* (Rio de Janeiro: Atlantica Editora, 1944); Lilia Moritz Schwarcz, *As barbas do Imperador* (São Paulo: Companhia das Letras, 1998); Lúcia P. Guimarães, *Debaixo da imediata proteção imperial: Instituto Histórico e Geográfico Brasileiro* (São Paulo: Annablume, 2011).

17 In 1827, for example, specially bound publications were sent to emperor D. Pedro I: letter, Saint-Hilaire to Adrien de Jussieu, Orléans, May 14, 1827. MNHN, ms 53.

18 Letter, Saint-Hilaire to Adrien de Jussieu, February 10, 1827. MNHN, ms. 53.

19 Saint-Hilaire, *Voyage aux sources du rio de S. Francisco,* vol. 1, xii–xiii.

20 *Catalogue méthodique des livres scientifiques légués à la Ville de Montpellier par Mr. Auguste de Saint-Hilaire, membre de l'Institut (...) 1855–1856.* Montpellier, Médiathèque centrale Émile Zola, ms. 322.

21 *Lettres inédites de Moquin-Tandon à Auguste de Saint-Hilaire* (Clermont-l'Hérault: Librairie Saturnin Létard, 1893), 11.

22 Letter, Saint-Hilaire to Visconde de São Leopoldo, Paris, November 7, 1839. Manuscript copy, Instituto Histórico e Geográfico Brasileiro (IHGB), DL 340.13.

23 *Lettres inédites de Moquin-Tandon à Auguste de Saint-Hilaire,* 113.

24 Alexandre Parent-Duchâtelet, *De la Prostitution dans la ville de Paris, considérée sous le rapport de l'hygiène publique, de la morale et de l'administration,* 2 vols. (Paris: J.-B. Baillière, 1836), vol. 1, 4.

25 João Baptista Calógeras, "Biographia. Manoel Theodoro d'Araújo Azambuja," *Revista Popular* (Rio de Janeiro) VII (1860): 171–181.

26 Letter, Saint-Hilaire to Ferdinand Denis, Vernet-les-Bains, July 24, 1847. Wellcome Library, ms. 7572.

27 Ibid.

28 Letter, Saint-Hilaire to Charles Naudin, Montpellier, June 24, 1847. Wellcome Library, ms. 7572.

29 Saint-Hilaire, *Voyage aux sources du rio de S. Francisco,* v. 2, 19.

30 Auguste de Saint-Hilaire, *Voyage dans les provinces de Rio de Janeiro et de Minas Geraes,* 2 vols. (Paris: Grimbert et Dorez, 1830), vol. 2, 38.

31 Letter, Saint Hilaire to Deleuze, December 4, 1816. MNHN, ms 2673.

32 Lorelai Kury, ed., *Comissão Científica do Império, 1859–1861* (Rio de Janeiro: Andrea Jakobsson Estúdio, 2009).

33 "Exame de algumas passagens de hum moderno viajante ao Brazil, e refutação de seus erros mais grosseiros, por hum Brazileiro," *O Patriota. Jornal litterario, politico, mercantil, &c. do Rio de Janeiro,* no. 3 (September 1813): 68–78.

34 Ilmar Rohloff de Mattos, *O Tempo Saquarema* (São Paulo: Hucitec, 1987); Kaori Kodama, "Os debates pelo fim do tráfico no periódico *O Philantropo* (1849–1852) e a formação do povo: doenças, raça e escravidão," *Revista Brasileira de História* 56, no. 28 (2008): 407–430.
35 Letter, Cambessèdes to Saint-Hilaire, Paris, October 25, 1827. MNHN, ms. CRY501.
36 Auguste de Saint-Hilaire, *Voyage dans le district des diamans et sur le litoral du Brésil*, 2 vols. (Paris: Gide, 1833), vol. 2, 151.
37 Sanjay Subrahmanyam, "Between a Rock and a Hard Place: Some Afterthoughts," in Simon Schaffer, Lissa Roberts, Kapil Raj, and James Delbourgo, eds., *The Brokered World: Go-Betweens and Global Intelligence, 1770–1820* (Sagamore Beach: Science History Publications, 2009): 429–440, here 439.
38 Saint-Hilaire, *Voyage à Rio-Grande do Sul (Brésil)*, 496.
39 Saint-Hilaire, *Voyage dans le district des diamans*, vol. 2, 271.
40 Saint-Hilaire, *Voyage dans les provinces de Rio de Janeiro et de Minas Geraes*, vol. 2, 349.
41 Renato Pinto Venancio, "Redes de compadrio em Vila Rica: um estudo de caso," in Carla Carvalho de Almeida and Mônica Ribeiro de Oliveira, eds., *Exercícios de Micro História* (Rio de Janeiro: Editora FGV, 2009): 239–261.
42 My goal here is not an in-depth analysis of the networks of trust that enabled Saint-Hilaire's travels. For an interesting approach to one traveler's fieldwork and the personal relationships he established, see Jane Camerini, "Wallace in the Field," in Henrika Kuklick and Robert E. Kohler, eds., *Science in the Field*, special issue of *Osiris* (2nd Series) 11 (1996): 44–65.
43 Letter, Saint-Hilaire to Muséum d'histoire naturelle, São Paulo, November 11, 1819. Archives Nationales, AJ15.
44 Letter, Saint-Hilaire to Charles Naudin, Montpellier, March 15, 1849. Wellcome Library, Ms. 7572.
45 Auguste de Saint-Hilaire, *Voyage dans les provinces de Saint-Paul et de Sainte-Catherine*, 2 vols. (Paris: Arthus Bertrand, 1851), vol. 1, 127.
46 All volumes of Saint-Hilaire's travel accounts contain descriptions of how climate influenced people in Brazil.
47 Saint-Hilaire, *Voyage dans les provinces de Saint-Paul*, vol. 1, 164 and *passim*.
48 Saint-Hilaire's travel accounts were ultimately published in four two-volume tomes, plus one posthumous book: *Voyage dans les provinces de Rio de Janeiro et de Minas Geraes* 1830; *Voyage dans le district des diamans et sur le litoral du Brésil*, 1833; *Voyage aux sources du rio de S. Francisco et dans la province de Goyaz*, 1847; *Voyage dans les provinces de Saint-Paul et de Sainte-Catherine*, 1851; and *Voyage à Rio-Grande do Sul (Brésil)*, 1887, posthumous.
49 Letter, Turpin to Saint-Hilaire, Paris, April 7, 1828. MNHN, ms. CRY 508.
50 On publishers and the book market in France, see Jean-Yves Mollier, *La lecture et ses publics à l'époque contemporaine: essais d'histoire culturelle* (Paris: PUF, 2001).
51 For an interesting discussion of the images and picturesque descriptions found in travel narratives, see Nancy Leys Stepan, *Picturing Tropical Nature* (London: Reaktion Books, 2001). On a more general level, see Anne Secord, "Botany on a Plate: Pleasure and the Power of Pictures in Promoting Early Nineteenth-Century Scientific Knowledge," *Isis* 93, no. 1 (2002): 28–57.
52 Alfred Moquin-Tandon, *Un naturaliste à Paris (1834)* (Chilly-Mazarin: Sciences en situation, 1999), 81.
53 Ana Maria de Moraes Belluzzo, ed., *O Brasil dos viajantes*, 2nd ed. (Rio de Janeiro/São Paulo: Metalivros/Objetiva, 1999).
54 Letter, Saint-Hilaire to Adrien de Jussieu, Orléans, [March 19], 1827. MNHN, ms. 53.
55 Saint-Hilaire's writing is akin to the early nineteenth-century descriptive statistics analyzed by Marie Noëlle Bourguet, *Déchiffrer la France. La statistique départementale à l'époque napoléonienne* (Paris: Ed. des archives contemporaines, 1989). But there

is also a visible influence from the methods of 1830s physicians and hygienists, like Alexandre Parent-Duchâtelet.

56 Letter, Saint-Hilaire to Charles Naudin, Montpellier, June 24, 1847. Wellcome Library, ms. 7572.
57 Letter, Saint-Hilaire to Charles Naudin. Montpellier, February 21, 1848. Wellcome Library, ms. 7572.
58 Brazilian emperor D. Pedro II helped fund the monumental *Flora brasilienis* (1840–1906), edited by Martius and finished by Ignatz Urban.

10 Commercial Statistics of Late Qing China between Global Interest and Local Irrelevance, 1860–1910

Stacie A. Kent

Introduction

For a significant period of time, spanning, roughly, the mid-eighteenth to the mid-twentieth century, intellectual and political projects around the world colluded in the creation and institutionalization of universals.[1] As a form of thought, universals structured thinking about reason, history, time, polity, and society, and as norms, universals authorized projects of state power, civil reform, scientific inquiry, and political revolution.[2] Brimming with content rooted in European historical experience, these universals more often than not cast the world "outside" Europe as a site of alterity, deviance, and the "other." Differences in epistemology, ways of life, and structures of power were explained in temporal terms as an absence of history or as an arrested development, or in spatialized terms such as environment, culture or religion.[3] Scholarship – at least in the humanities and some of the social sciences – no longer labors in this direction.[4] Instead, we study how universals are made (through politics, market forces, and awkward collaborations across difference, just to name a few culprits), work to recognize the specific contributions of non-European knowledge, and better appreciate the varied means through which human societies have organized to different ends.[5]

Yet, in some contexts universals still rule. As a small illustration, at present, it is difficult to find a single government that does not publish (often annually) official statistics, that is, quantitative data that make visible a wide range of material and social realities – how many people live within its official jurisdiction, how many tons of wheat or cocoa it produced in the past year, etc. The systematic compilation of such information is a relatively recent practice. Compilers of statistics will anxiously admit that in much of the world, until the twentieth century, no one collected the numerical data now so helpfully compiled into databases, such as ProQuest Statistical Insight or texts like *International Historical Statistics*. Social scientists in the early twentieth century had to decide that quantification was useful and desirable before routine, systematic data collection got off the ground. This is not to say that prior to the twentieth century states did not count. Only that they did so in conjunction with narrowly defined projects – gathering revenue, planning military campaigns, and maintaining state granaries. Today quantification is a universal method applied whenever it is thought that numbers, rather than words, will produce a more subtle, precise, and true truth.

DOI: 10.4324/9780367814540-12

What quantification enthusiasts can often forget (or at least not talk about very much) is that when we count, we do not so much record reality as *create* reality. With this insight in mind, the global take-up of state statistics is not merely a chapter in the universalization of social scientific method, but a site to study the creation of shared, universal realities. The engagements that (re)produced realities of population and gross domestic product are, unsurprisingly, varied. In this sense, there is no single or unified history of state-sponsored statistics. At the same time, however, the act of quantification, at the level of its form, works the same always and everywhere. It makes qualitatively different things (a woman's marriage and family, conditions at discrete workplaces) amenable to further operations of comparison, compilation, normalization, etc.[6] Quantification is a key means through which concrete social realities are gathered and homogenized for the purposes of defining problems, creating policies, and designing interventions.

We should not, however, mistake the universalizing potential of such social abstraction for the reality of its claims. Here I examine the deployment of commercial statistics in diplomatic negotiations between the Qing and British governments in 1867–1869 and the limits of their universality. These negotiations were the first time these two governments negotiated a peace-time treaty. In the preceding decades, the Qing and British Empires had fought two wars and used treaties to forge an awkward peace and new conditions for Euro-American trades in China.[7] As a result of these treaties, European and American nationals and the trade they engaged in became governed by rules and institutions apart from those governing Qing subjects and their trade. The treaties created a framework that both acknowledged Qing sovereignty and created possibilities for innumerable unsolicited interventions by foreign governments into matters of domestic administration, commercial, and otherwise. Commercial statistics were a product of these changes and a tool wielded by some to advance further changes in Qing commercial governance.

Enumerative Precedents

Enumeration and the centralized collection of numerical data were a long-standing part of China's imperial statecraft. The central government collected, for example, data on rain fall, supplies of state-stored grain and on the cost of labor employed on imperial construction projects. Yet, systematically collected data on commodity movements, their volumes, and values were unknown. Local gazetteers, one of the few systematic projects to catalog and describe foodstuffs and handicrafts, focused on their qualities and conditions of production. Even when the Qing state did begin to count new things (population, for instance) and to use new methods to compile and compare numerical data, commercial statistics in some ways remained on the periphery.[8] But why? One possibility is: if numbers create realities, there was something about the reality that commercial statistics created that did not resonate with Qing leaders. Certain forms of enumeration were readily assimilable into Qing statecraft. We should therefore study more closely the social abstraction commercial statistics produced and whether it communicated meaningful goals. The peripheral status of commercial statistics, moreover, provides an opportunity to query the historical specificity of the social abstraction they produced. Certain forms of abstraction *were* key to Qing governance. Officials educated in the classics were conversant in a cosmology that moved from concrete nature to abstract principles, and they were responsible for deploying those principles to generate order.[9] Something, then, was different about the abstraction produced by quantifying commodities.

A New Counting Operation: The Imperial Maritime Customs

Qing commercial statistics began in the 1860s with the quotidian operations of a novel agency, the Imperial Maritime Customs. The IMC, which officially commenced operations in 1861, ran a network of customs houses in China's treaty ports that policed trade and calculated tax payments due on ships and their cargoes.[10] Lead from 1863 to 1911 by Inspector General Robert Hart, the IMC was staffed at its highest levels by a coterie of Europeans and Americans, and its operations echoed new practices of British civil and military administration, including examination-based hiring, standardized accounting, and routine reporting protocols. It even ordered its stationary from London.[11]

Paperwork, it turned out, was central to new customs protocols and to the production of new statistical information. Under the IMC's regulations, foreign consignees and ship captains interacted directly with desks and clerks charged with processing goods into and out of China.[12] These interactions were mediated by papers that passed between the spheres of commerce and governance, recorded or archived by the IMC. They included consular reports of ship arrivals; ship manifests; applications to land and ship; certificates, memos, and receipts attesting to cargo and duties paid; registers of drawbacks and transit certificates; and return books.[13] Recorded in these papers were the ship's name and prior port of call, its country of registration and size, its cargo (items and quantities), its cargo's origins and values, and whether the goods were to continue their movement elsewhere. It is from these documents, transcribed and archived, that the IMC compiled its statistics.

Three enumeration projects were carefully attended to by the IMC's Inspector General. The first, called for by the IMC's supervisory body, the *Zongli geguo shiwu yamen* (Office of Foreign Affairs, hereafter Zongli Yamen), were quarterly revenue returns. In their initial form, which took shape between 1861 and 1863, revenue returns relayed three crucial pieces of information in English and Chinese. First, the amount of customs revenue collected by each office, second, the expenses incurred in collecting that revenue, and third, the amount of revenue allocated to pay off war indemnities.[14] The second project, initiated in 1863–1864, were annual returns to document "the real trade of the port, showing its value, with movement and consumption of goods." Each port was to report according to a standard form schedules of foreign trade, coast trade, shipping, duties, and goods of particular interest, such as opium. Unlike the revenue returns, which were a form of internal surveillance, the Trade Returns were to "be of real use to the public," by which Hart meant an Anglophone public, as they were not issued in any other language until Chinese returns commenced in 1889.[15] Though of questionable utility for merchants, the Returns did approximate a national accounting of China's participation in global commodity flows and capital reproduction.[16] The third project, begun in 1865, complimented the returns with "correct" and "interesting" narratives that contextualized numerical data. These annual reports, also available only in English until 1889, were gradually standardized over time and culminated in the Decennial Report series, initiated in 1891, which surveyed for each port 26 categories of information, including commerce, geography, political disturbances, the opium trade, local money markets, and development projects (rail, telegraphy), and provided ample tables of trade volumes, values and revenues.[17]

With these projects the IMC took on the responsibility to render transparent as possible the contributions of global commerce to Qing fiscal health as well as the movements, commodities, values, and volumes that constituted that commerce. Yet, these two tasks were incommensurate. Revenue figures were key to Qing statecraft as the empire attempted to pay off its international obligations, battle internal rebellion, rebuild, and develop new military capacities. The latter information, which captured in numerical form scales of production and consumption, as well as the thickening or thinning of commercial integration within the Empire and between the Empire and other parts of the world, was written not for Qing authorities, but for a Euro-American public that recognized commodity volumes and values as metrics for social well-being. This information circulated broadly outside of China to other Asian ports of global trade, such as Singapore and Yokohama, as well as to British commercial centers, where IMC statistics populated government reports and became items of breaking news, particularly in discussions of pressing national interest such as the silk, tea, and opium trades.[18] For all their popularity abroad, however, there is little evidence Qing authorities thought their ability to govern was improved by knowing the value of cotton shirting that entered the country.

Treaty Revision and Social Order

> It is, of course, impossible to talk about the economy of nation by looking at 'it.' The 'it' is plainly invisible, as long as cohorts of enquirers and inspectors have not filled in long questionnaires […] only at the end can the economy be made visible inside piles of charts and lists.
>
> (Bruno Latour, "Drawing Things Together")

Scholars have observed, and even the amateur baseball statistician knows, that quantification is basically a tool for problem solving, for making judgments, for deciding courses of action. When is it the right tool for the job? Or alternatively, for whom is it the right tool? When given the opportunity in 1867–1868 to use newly crunched IMC data to revise the tariff on foreign goods and ease British pressure for wholesale fiscal-administrative restructuring, the agency that led Qing treaty administration, the Zongli Yamen, passed. Not because they did not use numbers to make decisions, not because they mistrusted the calculations, and not because they could not think abstractly. Rather, they passed because the problem the numbers identified and solved was not theirs. While commerce was important to the social welfare of the empire, the Qing government addressed itself to the particular qualities of this activity rather than general measures that posited an abstract equivalence between different forms of social activity.

In the epigraph that began this section, Bruno Latour is half right. The four principal indicators used by the U.S. government to render visible the U.S. economy – gross domestic product, personal income, trade balance and account balance – are all quantitative fabula created by the roughly 500 economists and administrators at the U.S. Bureau of Economic Analysis. The enumeration, though, is really a last step, a final product and instrumental short hand that follows from changing social relations that have generated complex webs of interdependence. Consider, for example, the following passage written by British merchants at Chefoo (Yantai) in May 1867:

> The importance of Chefoo as a central depot for the trade of the north of China is daily becoming more manifest by the qualities of merchandize imported here from the other ports in the Gulf of Pecheli (Bohai Sea) for transmission to the southern provinces of the Empire, and by the distribution from here of British manufactures through this and neighboring provinces. We shall see in the present year […] a fresh impetus and development to our trade. […] As soon as the projected railroad from here to Tsianfoo (Xi'an) is constructed, Chefoo will become more and more the maritime door of the north of China, and having […] a direct import trade with England, and a railroad to the interior, will enable it to supply the markets of Shantung (Shandong), Honan (Hunan), and Shansi (Shanxi), at all times with British manufactures. […] Good coal can be obtained at a place only twenty miles distant from this, and if permission could be got to work these mines, and to construct a tramroad in order to bring the coal here, great good would result in many ways […] especially if the facilities such a supply of coal would afford to steamers, both of the Royal and mercantile marine.[19]

In their optimistic report of the commercial potential of Chefoo, the memorialists paint a picture of what we might now call an economy – an interactive organization of production, circulation, and consumption of goods and services. One arm of this economy stretches from Manchester to north-east China, another reaches up and down the coast, and a third pushes inland to a hinterland waiting to be incorporated into global circuits of goods and capital. Interactions between British producers, merchants, and Chinese consumers have stimulated and in turn will be stimulated by capital flowing into new spheres of investment, production, and consumption, in particular, railroad construction and coal mining. Coal fields in Lancashire will conjoin with coal fields in Shandong in a symbiotic circuit.

Although the memorialists do not give any quantitative measure of the scale of these activities, we still have a picture of an economy and forms of social abstraction. The memorialists do not bother to discuss the kinds of merchandize that are moving, what matters is that "quantities" are. The social good is not a concrete finite end serviced by the qualities of certain goods but an increasing scale and geography of entangled production, circulation, and consumption, that will hopefully in the near future take place more quickly. This vision of an economy, perhaps unsurprisingly, is also a vision of capital: values perpetually moving, exchanging, and growing with no particular end in sight. But, the memorialists continued, and many others throughout China concurred, the Qing government posed serious threats to capital so understood.

Whatever hopes and optimism attended two rounds of treaty making between the British and Qing governments, in the mid-1860s, very few on either were satisfied with the realities that would prevail in the decades thereafter. Chinese merchants and shipowners petitioned about lost livelihoods.[20] Qing officials throughout the empire criticized unscrupulous merchants (Chinese and non-Chinese) who colluded to avoid taxes and smuggle goods.[21] British merchants lodged complaints against illegal taxation and unwarranted confiscations by ignorant, over-zealous, and corrupt officials, and they agitated for steamships on inland waters, railroad construction, and mining concessions. The British Minister in Beijing, Rutherford Alcock (in office 1865–1869), connected commercial difficulties to multiple illnesses endemic to the empire. In a memorandum to ministerial colleagues in Beijing written in November 1867, Alcock pointed to the need for "fundamental changes" in Qing halls of government if treaties were to be reliably executed. "The greatest obstacle to the execution of the most important stipulations for the protection of life and property and the free development of commerce, arises from internal disorder and misrule in the provinces," he wrote. "These again are to be traced partly to defects in the Government at Peking, and partly to the venality and incapacity of the whole body of officials throughout the empire."[22] For Alcock, who had received numerous mercantile memorandums on commercial conditions at Chinese treaty ports and who had negotiated numerous individual merchant claims with counterparts at the Zongli Yamen, all the individual complaints and discrete events added up to system-level defects. The design of government in China – of delegated powers, of taxation, of appointment, of funding – was all wrong.

In the fall of 1867, Alcock knew this diagnosis brought him to the edge of calling for what we might today refer to as "regime change." And he offered foreign trade as justification for doing so: "the foreign policy of the country is too intimately connected with its internal administration and constitution to allow of their separation... the domestic policy of China and the whole internal administration become matters of legitimate interest to Treaty Powers having large material interests at stake."[23] If how China conducted its internal affairs was going to shape the possibilities for British capital circulation and reproduction, then the British government had a right to intervene in Qing internal affairs. Alcock, however, was wary of using overt coercion. The best chance to improve the situation, he assessed, was the upcoming treaty revision window. Stipulations within the 1858 Treaty of Tianjin called for the possibility of revision after ten years, and Alcock, with London's concurrence, felt revision was an opportunity to see what further negotiations could achieve.

Politics of the Concrete

Notified that the British were interested in revisiting the Treaty of Tianjin, the Zongli Yamen asked for provincial input into how it might proceed. In the winter of 1867–1868, Qing officials did not share Alcock's grand visions for what treaty revision might accomplish. Instead they saw yet more lists of British requests with unfavorable concrete repercussions for the polity and its people. History, moreover, had shown that Europeans and Americans were quarrelsome and competitive. China needed to resist being bullied.[24] In one memorial to the throne obtained by the British Embassy, a prominent governor-general, Zeng Guofan, warned,

> Foreigners in the east and west for several hundred years have been making and un-making kingdoms, each kingdom wishing to deprive its neighbors' subjects of some advantage, with the hope that its own subjects will ultimately profit thereby […] they wish to damage our merchants […] [who] suffer in mute agony, and will be driven to extremity. If trade in salt is conceded to foreigners, salt merchants will suffer in business; if the building of godowns [in the interior], the establishments already existing will suffer; if small steamers be allowed, in the interior, native craft of every size, sailors, and pilots will suffer; if they are allowed to construct telegraphs and railroads [owners of] carts, mules, chairs, and inns, and the coolies' livelihood will suffer.[25]

For Zeng, experience had shown that foreigners were a greedy, selfish lot, happy to take their own benefits at the expense of others. As they swarmed in China, looking for advantages, they threatened particular groups of Chinese people, from merchants to day-laborers to sailors and inn-keepers.

Zeng's concerns were shared. His colleague Li Hongzhang also characterized the British and other foreign countries as abstruse and coercive. Their demands, if not firmly countered and blunted, threatened the ability of the people to nourish themselves and the ability of the state to fund itself.[26] At the Zongli Yamen, Imperial Prince Yixin echoed provincial authorities' irritation that the British did nothing but seek profit and pressed again for further concessions (of the sort listed by Zeng). In 1863, Yixin had already lost a protracted contest with the British Embassy to save Chinese soybean shippers from mass unemployment as foreign steamships took over the trade. Now, Yixin felt many of the additional British asks were to be resolutely refused on the grounds that their concrete consequences mattered: they either concerned government administration or granting them would obstruct people's livelihoods.[27] Furthermore, it wasn't evident to the Prince what China gained from the treaty relationship with the British government. He argued, "The Chinese government does not issue orders in matters of trade relations. Our produce is well-provided for, and really we have no need to avail ourselves of foreign countries."[28] While earlier Sinologists have argued that such sentiments reflect the anti-commercial animus of the Qing, I think this statement and others do quite different work. Namely, they help us understand why the commerce pursued by the British struck this highly commercialized polity as so alien.

First, it is crucial that Qing officials like Zeng Guofan did not foreground growth as the goal or animus of commerce. When Zeng rejected British requests for steamships, coal, telegraphs, and railroads, it was not because of the technologies, per se, but because those technologies would take away the livelihood of Chinese persons. Central to the thrust of Zeng's concern was the view that trade was a zero-sum competition in which advantages accrued to particular groups, and groups with a greater advantage would cause others to suffer. How could speed-up be desirable if it also produced massive social disturbance? A second reason Qing leaders hesitated to adopt the British agenda was that they didn't see how the quicker, larger-scale circulation of British stuff would benefit the Empire. If commerce was the circulation of useful things, the Chinese had the useful things they needed. At the same time, however, Yixin keenly perceived how the British were different: they needed China to fulfill their desire for profit, a qualitatively different matter in so far as it had no finite end or discrete object.

The differences between the social order in China and the social order in Great Britain, which was being proposed to China as a model for governance, set up the framework for negotiations. From the Qing point of view, the imbalance between Chinese needs (use-values) and British ones (value) put China in a situation to be magnanimous. It could, Yixin argued, without any harm to tax revenues, lower some tariff duties, open additional ports, and make new arrangements for tax refunds.[29] These gestures would provide some of the profit sought by British merchants. At the same time, given how assertive (and belligerent) Britain had been, the Yamen did not want to be too quiescent and, drawing from the advice of officials like Li Hongzhang, advanced its own list of changes to negotiate.[30] The goals of these changes, the Yamen explained, were to "prevent foreign merchants from monopolizing goods traded by Chinese merchants" and to prevent a popular form of tax evasion in which "foreign merchants secretly give a foreign flag to a Chinese boat." The Yamen's wish list also included increased taxes on tea, silk, and opium and new arrangements to guarantee the payment of transit taxes. Both sides should benefit from the revision, Gong reasoned.[31] Benefits for China would be employed Chinese merchants and increased government revenues, something that would become quite tangible as the state repaired damages from the Taiping Rebellion and produced new armaments.

With their two wish lists, British and Qing negotiators not only put forth proposals the other side would resolutely refuse, but each list conceptualized the issues differently. Alcock forwarded changes – including a restructured fiscal system, more waterways for steamships to ply, and more permanent foreign presence outside of treaty ports – intended to generate quicker, larger, and more reliable capital circulation. Qing officials considered how government decisions could shape the conditions under which particular groups of persons pursued their livelihoods. By all appearances, many would have preferred to leave things as they were. Nonetheless, pressed into negotiations, Qing authorities focused their attention on the most concrete benefits they believed revision could bring: less smuggling and more revenue. The treaty negotiations were, in the end, a debate about what social good the Qing state should support: the ability to protect people and revenue or the general process of capital movement and growth as a means to produce a certain kind of social good. The two Empires stood on either side of a large gulf.

Counting and Calculating: Social Abstraction in Negotiations

As negotiations continued into the summer of 1868, it seemed like numbers might build a bridge between the two sides. Emerging at the center of Alcock's revision negotiation was an issue fundamentally internal to Qing administration – its domestic tax system. As a matter of imperial design, provinces practiced some fiscal autonomy, particularly with reference to an emergency levy on commerce, referred to as lijin. Lijin began in 1854 amidst the Taiping Rebellion, and despite its ostensibly temporary nature, it outlasted the dynasty. British treaty negotiators in 1858, aware of lijin if not in name then form, had removed British goods from the reach of Qing "transit duties" by commuting all transit taxes into an additional 2.5 percent tariff charge levied at the time of import.[32] The resultant "transit system," de facto administered by the IMC and provincial authorities, was plagued by administrative inconsistencies, merchant collusion, and plain old confusion. One result of its troubled implementation was plausible claims of illegal extra-treaty taxation.[33] Convinced of the truth of such claims, Alcock concluded that inland taxation was the most significant threat to British capital. "The trade is not only throttled, so to speak, by the leking [lijin) and local taxes levied at the ports, but the maintenance of the right to impose them involved a principles which... defeats the main object of the Treaty," he complained.[34] The point of a commercial treaty, Alcock argued, was to place absolute limits on how much tax the Chinese government might levy on foreign goods, and to allow provincial authorities unilateral authority to tax them, during transit or at the time of sale, was tantamount to locking the wheels and obstructing the axle of the British commercial wagon.[35]

Alcock's vision of the inland taxation problem shared with the Chefoo memorialists' key forms of social abstraction. To begin, he was preoccupied with what we might call the total circuit of capital. Goods manufactured in Great Britain found their ultimate sale in China. Likewise, Chinese produce ultimately arrived in the teapots of the men and women of England. This trans-oceanic, trans-imperial movement meant that taxes levied on one end of the world affected what could be produced and consumed on the other. With the health and growth of commerce in mind, heavy and frequent taxation within Qing territory was "irresponsible" whether levied on foreign imports or on Chinese goods intended for export. Notably, in discussions with the Yamen, Alcock never singled out any particular goods and their use. He never said, for instance, that British cotton goods, sold cheaply in China, would better clothe the Chinese peasant. Nor did he say that Chinese tea, sold more cheaply in London, would improve the livelihood of Chinese farmers. Rather he repeatedly invoked "trade" as the thing that treaties protected and Qing tax policy was obliged to tend to. The difference between cotton shirtings, tea, and trade, of course, is that the former two have use-values immediate to their consumption. The use-value of trade is to generate more trade. In short, Alcock's complaints pursued a project wherein the Qing state would stop governing the movement of useful goods and start governing movement itself.

With movement in mind, Alcock went further to suggest an even larger change in Qing statecraft: the subordination of tax policy to capital reproduction. As he made clear to Yamen representative Robert Hart, within the framework of treaty relations, Qing governance subordinated its political rights to the needs of commerce:

> Of what possible value can it be to Great Britain or trade to have the carefully guarded treaty right of effecting an entrance into Chinese ports for their goods, and even a right of transit across the whole territory to any given point of sale and consumption, on payment of a certain moderate, fixed, and predetermined duties, if behind this was reserved the power in Chinese hands to surcharge such goods with taxes rendering both sale and consumption impossible among the natives? [...] Granted a territorial and municipal right of taxation on all trade, it must not be wrongfully exercised.[36]

Treaties, Alcock argued, had authorized a project of trans-imperial capital reproduction in which Chinese consumers were the end of the capital circuit – the point where reflux began. By authorizing this integration, the Qing government had pledged itself to create conditions that, if not hospitable, were certainly not inhospitable.

From Abstraction to Enumeration

While Alcock was focused on the incidence of inland taxation, Qing negotiators were focused on the question of tax rates on key goods. Early in negotiations they had granted lower rates on items the British felt bore disproportionate tax burdens and which the Yamen knew generated little revenue. These included cotton and woolen goods, watches, tin places, timber, and black and white pepper. In return, the Yamen made two counter-proposals. The first was to aid imperial revenue by doubling tariff duties on tea and silk, the two largest foreign exports. The second was to level the playing field for merchants from all countries by re-equalizing all tariff rates to five percent ad valorem and rectify actual tariff practice with tariff principle.[37]

The British Minister was interested, but also wary. His foremost concern was to ensure that British circulated goods could be sold, and preferably to more people in more places. This meant, any solution that would rid British trade of "evil" lijin taxes, the "incubus" that throttled commerce was on the table.[38] Was it possible to exchange increased rates on tea and silk for the end of lijin? Yet, he also anticipated that increased taxes on tea and silk would greatly upset foreign merchants and incur the opposition of their governments. He was also less persuaded equalization was a good idea, as the tariff had been crafted to privileged the specific trade interests of each treaty signatory.[39] Rather than worry about how much money any item of trade could generate, instead, Alcock argued, the government should consider overall trade volumes. "When trading is robust, tax revenue will increase," he wrote to Yixin, offering steamships, rail, and coal as means to expand trade.[40] Stated somewhat differently, the quantities that mattered for governance, Alcock argued, were not specific numbers of individual commodities, but the number of any and all goods as a measure of overall social activity. Wiser political economy, the British Minister urged, would grow the absolute volume of circulation.

Alcock's argument, if it did not sound strange, certainly sat awkwardly alongside imperial precedent and practice. The Qing Empire had limited "production" of pearls and furs to certain clans based on the political ties to the dynasty, had forcefully argued to protect the employment of ships that circulated soybeans, and had routinely offered tax exemptions on strategically important goods (tribute rice, timber). In other words, its governance had worked over a heterogeneous landscape populated with non-interchangeable goods, occupations, and relationships.[41] And while counting pearls and pelts, ships and sailors, and calculating tax rates were key to managing these fields of activity, the numbers served to describe the particular, not, as would become statistical practice later, to create a "large number… that is not made of instances."[42]

At this point, in the summer of 1868, negotiations took a strange enumerative detour. On one hand the detour was an outcome of what common ground the two sides shared – a willingness to discuss how much tax was levied from foreign trade. But, despite taking place on shared ground, it produced no policy or course of action. Suggestions that emerged from counting up volumes and values were disregarded by all almost as soon as they surfaced. As such, the event offers a limiting case for the universality of universals and illuminates further the historical differences that generated both global interest and local irrelevance. The movement of useful goods was not to be confused with a sphere of aggregate commercial activity.

The enumerative detour took place in private side communications between the British Minister and Robert Hart, who as Customs Inspector General was a key Qing negotiator. On August 7, shortly after the Yamen proposed tariff equalization, Hart wrote to Alcock with several sets of calculations gleaned from the IMC's statistical returns. One set enumerated the duty collected in 1867 on tea, silk, and "all other exports." Another set enumerated the same for opium, cotton manufactures, woolens, metals, and "all other imports." Together, these two tables established a baseline figure for Qing customs revenue: 7.9 million taels. Hart then calculated several hypotheticals based on current market volumes and values: 1) what if duties on tea and silk were doubled, and all other duties, excepting opium, were lowered to 2.5 percent ad valorem? The result: an increase in revenue to 10,160,000 taels. 2) Were there articles of trade that produced such little revenue they did not pay the "costs and trouble of collection"? Setting a threshold of 1,000 taels, the answer was yes: 480 items of export and 400 imports. If the government did not collect duties on these items, Hart calculated, it would only lose 200,000 taels in revenue, annually. Next, Hart listed 12 imports and 20 exports that could produce 15 million taels revenue at certain rates. Key to this increase were higher duties on silk, tea, and opium. Finally, he asked the question, if those same imports and exports were taxed without increased duties on silk, tea, and opium, what would be the result? Answer: a loss of 445,000 taels in revenue.[43]

Using IMC data on how many goods were crossing Qing borders and their current market values, Hart's calculations established that the majority of Qing revenue was generated from a small set of goods, foremost silk, tea, and opium. Alcock responded to the calculations with gratitude: "your statistics," he wrote to Hart, "are just what I wanted, and invaluable. They go far to confirm some general ideas I have for some time been turning over in my mind. For any effective revision, it is obvious that the whole question of the Customs dues, maritime and transit, should be dealt with comprehensively."[44] What was going on here? What were the numbers in service of?

Hart and Alcock, despite being on opposite sides of the negotiating table, held the same general opinion about Qing fiscal arrangements. Hart, like Alcock, was no fan of Qing political economy. In one of their earliest private letters during negotiations, Hart referred to "the peculiar twisted way of looking at things peculiar to the Chinese official mind."[45] Yet, he positioned himself as a go-between who understood that mind and could press upon it more enlightened statecraft, which in this instance would generate needed revenue by different means. In a lengthy letter to Alcock relaying the Zongli Yamen's position on British requests, he conscientiously laid out the agency's reasoning, but also assured Alcock that he "advocated your views." "What I write now," penned Hart, "is opposed to what I argued for."[46] Hart understood the questions had two sides, but he was predisposed to favor the British view.

If the British goal in treaty revision was to secure conditions to grow trade, the enumerative detour redefined the problem of capital growth as a problem of tax rates. In doing so, the detour proposed a new calculus for Qing statecraft. Rates were the one item Chinese officials were eager to discuss, as the government, burdened with indemnity payments, reconstruction, and arms production, could not sustain itself with any less revenue and surely could use more. Hart's calculations attempted to reconcile the Qing revenue project with British efforts to grow commercial activity. Could lijin be eliminated and the Qing Empire not suffer? What would need to be the new balance of taxes to support both projects? Could equalization to five percent generate enough revenue? Was it necessary to increase rates on some goods? These questions, as well as the calculations to answer them, were novel within Qing statecraft, which typically prioritized two factors in its commercial revenue considerations: the ability of affected merchants to pay without harm and the circulating good's use. Raising taxes on silk, for example, was attractive because it was a luxury good, and its trade was robust. Hart's numbers, however, pointed toward an alternative decision-making calculus. He set aside both the use-value of the goods and the social organization of commerce, focusing instead on the quantitative dimensions (how many, how much value) of commodity movements. On this basis alone, he compared goods as potential revenue generators, and reintroduced goods to Qing authorities as interchangeable means to engineer a pre-established revenue goal. Balancing numbers could produce a wiser, growth-oriented political economy.

Alcock responded to Hart's calculations with his own. He first considered consolidation and simplification. Eliminating 900 items from the tariff, Alcock concurred, would produce revenue losses of 200,000 taels, "something altogether insignificant, even if not covered by saving in collection." Further simplification to a tariff on 11 exports, cottons and woolens, metals, and opium would produce a much greater revenue loss: 1.31 million taels, including losses in transit taxes. But, Alcock was optimistic that readjusting rates on large-volume "staple" articles of trade would more than compensate for the losses. Doubling the tariff duty on opium would alone generate an additional 2 million taels. Doubling the duty on silk to bring it back to five percent ad valorem would increase revenue by 330,000 taels. Having hypothetically generated 1.2 million more taels for the Qing government, Alcock then gave British merchants a tax cut on tea. This tax cut, he admitted, absorbed the newly generated surplus, but it caused no losses and promised greater future revenues "from the development of trade under lighter burdens." Alcock summarized his proposals as a "vast simplification" that would economize on Customs administration, equalize duties to a light rate of five percent, and reduce duties on the one item (tea) "most susceptible to indefinite expansion," and through the elimination of lijin "release" foreign trade from the "heaviest and most injurious burdens ever laid upon it." "This," he wrote to Hart, "is the revision of Tariff I would propose as certain to be effective in removing many causes of dissatisfaction and difficulty, and likely to prove beneficial to both sides."[47] Although their proposals differed in their particulars, the two British negotiators were equally enamored with the alternative fiscal futures switching out numbers produced.

Why Numbers Didn't Matter

If tariff simplification seemed to offer common ground for the British Embassy and Zongli Yamen, its eventual fate highlights the deeper differences between the two sides. Hart and Alcock were both enthusiastic about creating a streamlined tariff. Alcock projected that "by the simple process of readjustment" not only could the government secure its revenue, but British trade could also be released from local taxation, an "irresponsible and uncontrolled" practice, that "may at any moment become a source of pressing danger."[48] Hart was enamored with the potential simplicity of collection. In its 1858 form, the tariff enumerated hundreds of classes and subclasses of items, many with set duties that reflected market value at the time the tariff was set. "The (1858) Chinese Treaty Tariff," Hart wrote to Alcock, "is admirable in that it presses heavily on no one article; but from the higher point of view, it is, to my mind, the most absurd thing of the sort in existence."[49] Both sides furthermore agreed that silk, tea, and opium were most fiscally generative, and Hart assured the British Minister the "Chinese mind" understood the logic of numbers. But, once the numbers were calculated, they held different meanings for each polity. For the Yamen, these items' status as luxury consumption meant taxes could be increased without causing general poverty.[50] The British Minister could not be so sure. He knew that raising rates would diminish the volumes and profit margins of trade, and putting trade at risk was tantamount to threatening continued growth and prosperity for China and Great Britain alike. The common ground was more apparent than real; tax rates did not so much bridge differences in how each polity functioned, but highlighted them.

To see these differences, it helps read the revision debate through two questions about governance: would good governance approach commodities as use-values or exchange values? Would good governance pursue the self-interest of merchants or balance it with other considerations? Of course, we now know there is no universal good governance, but in China in 1868, only one side argued such. While Alcock pushed that facilities for trade were "the mother of revenue," Qing officials were sensitive to spatial particularity and the ill fit between concrete realities in China and what a purely quantitative view made possible.[51] As negotiations reached critical mass in December 1868, Yixin offered a counter-lesson to Alcock in what it meant to govern. Merchants, Yixin explained, "do not reflect upon circumstances of time or place, or the practicability of proposed measures [...] their sole object being gain." But governments could, like a physician with a patient, act wisely to ascertain what was appropriate to the whole polity at a particular time and in a particular place. He warned the British Minister, "It may happen that the drug which is a specific in one case, will be valueless or even hurtful in another."[52]

Yixin correctly perceived that underlying the British Embassy's proposals was disregard for what we might now call the uneven effects of capital: what might benefit the aggregate might also cause pain to the particular and distributions of benefits could be geographically unbalanced. This imbalance is a contradiction internal to capitalist society in which the production of useful things is subordinate to the production of capital itself.[53] Witness, for example, the massive amounts of waste produced by "fast fashion" or wealth and public service disparities between San Francisco and Detroit. The revision negotiations translated this contradiction endemic to capitalist society into the sphere of geopolitics, where Qing officials drew attention to the implied subsumption of governance to the logics of capital reproduction. In a remarkably similar debate eight years later, Li Hongzhang would directly indict Hart, "if we are to collect customs tax, we must collect customs tax on all cargos... If we only consider the profit motives of foreign merchants, then where is the body politic (*zheng ti*)?"[54] More than just an invocation of sovereignty, Li faulted Hart for failing to distinguish the polity from the pursuit of commercial growth.

If principle hardened Qing officials against British claims they should govern for movement and aggregate quantities, qualities of the Qing Empire's commercial and fiscal structures queried the wisdom of plans that originated in statistical considerations. Hart gave Alcock a terse one-line explanation of the fate of the simplification proposal: "[Qing negotiator] Wen Xiang likes the idea of simplification, but thinks the time has not yet arrived."[55] Two difficulties mitigated against the plan. First, the British cause in China – a consistent empire-wide tariff, changed customs procedures and treaties to secure privileges of travel and certain conditions of exchange – had resulted in a parallel commercial-regulatory network that intersected with and competed with that produced by Chinese merchants and Chinese junks, which were regulated by different rules and non-IMC customs houses. Within that bifurcated regulatory regime, to abolish duties on 900 articles of trade when they were carried by British vessels or on British account through the IMC would put Chinese traders at a competitive disadvantage and induce them to smuggle.[56] Until all trade was brought under the administration of one institution, it was not time to simplify the tariff.[57] There is not a little irony here that the treaties' earlier arrangements to grow commerce stymied later efforts to advance that project.

This institutional difficulty belied deeper incongruences between the wisdom of numbers and extant Qing governance. Simplification of the import/export tariff posed an unenviable range of choices for the Qing court: concede an upper hand to foreign merchants who would benefit from lower levies or eliminate the same taxes as they were collected on Chinese merchants at customs houses, lijin transit barriers, and points of retail sale. The latter course would not only lead to revenue losses not anticipated by Alcock or Hart, who confined their calculations to the import/export trade, but also implied fiscal centralization that would rearrange the balance of power between Beijing and the provinces. In other words, to simplify the import/export tariff was not merely a matter of foreign trade but would mean altering how the government raised revenue from internal circulation as well. Moreover, given that much of that work was performed by provincial and local governments, changing the import/export tariff would have undercut the extant delegation of the power and responsibility to govern. For an imperial order that relied on geographically dispersed power and money, the immediate wisdom and feasibility of converting the hundreds of revenue streams represented in the tariff, which filtered through the empire, into a few rivers all directed to Beijing were less than obvious.

The Revision that Wasn't and Future Enumeration

In its final form, the "Supplementary Convention to the Treaty of Tianjin," signed in October 1869, did not include a simplified tariff, which proved to be little more than an intellectual exercise and diplomatic fantasy. Instead, the agreement gave each side some of what it wanted, without explicitly challenging the underlying political economies. Alcock procured tax arrangements intended to cheapen and secure circulation of British-imported cotton and woolens as well as Chinese produced (e.g., tea) exported abroad. The Yamen won higher tax rates on silk and opium as well as tax arrangements that placed Chinese and foreign merchants on more equal footing and redirected some commercial revenues into provincial treasuries.[58] For the two British men who had worked so hard to create an alternative taxation path for the Empire, failure to simplify the tariff went without remark, and they turned instead to the future significance of the agreement, which was not merely its immediate provisions but the steps it took toward the fuller realization of British ambitions. The "concessions" gained, Alcock pressed upon the outgoing Foreign Secretary, provided "a wedge for introducing and developing something better," an achievement that much more notable in light of "the ineptitude and ignorance, as well as the corruption and misgovernment which everywhere prevails in China."[59] Yet, it was precisely this future realization that the commercial and manufacturing communities in Great Britain did not see, and on the basis of their complaint that the revision did not do enough, the British government refused to ratify the agreement.[60] Better to endure the present difficulties than institutionalize limits on horizons of British trade.[61]

So much for the limitless project of limitless growth. What about the statistics? Did the negotiations begin a new way of thinking about fiscal policy in China? Taking a step back, aggregative enumeration of volumes, values, and revenues offered in this moment both a way to conceptualize social production and consumption as a totality and a tool to re-engineer fiscal practice in service of that totality. Read in this light, the failure of Alcock and Hart to persuade the Qing that better governance would result from such political-economic abstraction and fiscal centralization constitutes a moment when a universal model hit a limit. On one hand the limit was practical and reflected the institutional arrangements of government. On the other hand, there was no take-up because there was, from the Qing point of view, no corresponding reality. The Qing governed a polity of silk merchants and indigo dealers and many others whose circumstances of social reproduction were particular, not general.[62] The calculations mobilized by Hart and Alcock, however, stemmed from a political economy that specifically promoted general growth, even at the cost to some.

Indeed, even several decades on, at the beginning of the twentieth century when statistics were of increasing interest and use in China, the relations of production and circulation that the Qing governed continued to be cast in terms of concrete particulars, rather than abstract aggregations. In 1908 a new Bureau of Agriculture, Industry, and Commerce generated its first set of commercial statistics.[63] They looked little like the IMC's. Where the IMC listed classes of commodities, their volumes, and values, these statistics cataloged names of firms and individuals, their locations, activities, and sources of capital. In short, these statistical operations focused on firms not goods; they refused to reduce either economy or polity to commodity flows. As a final clue as to how the Qing Empire took stock of its resources in terms of discrete human activities, the 1908 report marks the distance between indigenous forms of knowledge and the IMC's statistical projects. The latter, with their focus on volumes and values of commodity movements, attended to aggregate and abstract effects of human activities. The Qing preferred to measure their strength in people located in places, a way of knowing the world increasingly submerged over time under cascades of macro-economic data, in China and around the world.

Notes

1 While many cultural and epistemological traditions employ universals, of interest here are those associated with the projection of European power and capitalism. These are often traced to the Enlightenment and thinkers ranging from Adam Smith to Karl Marx to Immanuel Kant, though articulations – particularly in the field of economics – continued into the twentieth century. See for example W. W. Rostow, *The Stages of Economic Growth: A Non-Communist Manifesto* (Cambridge: Cambridge University Press, 1960). The study of the Enlightenment and its universals spans subfields of intellectual, political, and cultural history, as well as the history of science. As a historian of China and capitalism, my entry into the presence of these universals has been through their appearance in eighteenth-century social theory (e.g., Adam Smith's *The Wealth of Nations*),

China studies, and through their critics (see notes below). On the global dimensions of Enlightenment thought and of the creation of universals, see Sebastian Conrad, "Enlightenment in Global History: A Historiographical Critique," *The American Historical Review* 117 (2012): 999–1027.

2 Recent work on the creation of universal forms of thought within Europe includes Frank Palmeri, *State of Nature, Stages of Society: Enlightenment Conjectural History and Modern Social Discourse* (New York: Columbia University Press, 2016). Representative practical projects can be found in Edward Said, *Orientalism* (New York, NY: Pantheon Books, 1978); Michel Foucault, *Discipline and Punish: The Birth of the Prison*, trans. Alan Sheridan (New York, NY: Vintage Books, 1977); Timothy Mitchell, *Rule of Experts: Egypt, Techno-politics, Modernity* (Berkeley, CA: University of California Press, 2002); Partha Chatterjee, *The Black Hole of Empire: History of a Global Practice of Power* (Princeton, NJ: Princeton University Press, 2012); *Nature and Empire: Science and the Colonial Enterprise*, ed. Roy MacLeod, *Osiris* 15 (2000). While colonial science may have instantiated the universal aspirations of the European scientific enterprise, it also depended on and integrated local, indigenous knowledge, sites, and actors.

3 J. M. Blaut, *The Colonizer's Model of the World: Geographical Diffusionism and Eurocentric History* (New York, NY: Guilford Press, 1993). In certain areas of thought, such as social evolution and the emergent discipline of anthropology, time itself was spatialized. See Johannes Fabian, *Time and the Other* (New York, NY: Columbia University Press, 2000 [1983]). From the mid-eighteenth century forward, China was among the casualties of new forms of universalizing historical thought, which in one variant cited human practice as potential impediment to natural progress and in another posited the political modernity of Europe as the latest (last) stage in universal human development. By the metrics employed in Europeans' reflection on their own recent past, at the dawn of the nineteenth century China was lacking in commodities, technologies, intellectual vivacity, and distance from "church, ritual, ceremony, superstition, and magic, and from custom and habit." See James Hevia, *Cherishing Men From Afar: Qing Guest Ritual and the Macartney Embassy of 1793* (Durham, NC: Duke University Press, 1995), 70–72. These indictments were later re-worked in John King Fairbank's seminal *Trade and Diplomacy on the China Coast: The Opening of the Treaty Ports, 1842–1854* (Stanford, CA: Stanford University Press, 1963), which argued the "essential nature" of the Chinese state could not cope with "irreversible trends of Western history, manifest in the growth of science and technology, trade and industry, nationalism and the modern state." Chief among these essential characteristics were duplicity, stagnation, blindness to "economic realities," and a preference for ritual and prestige over the material benefits of trade (83). Confucianism, in particular, was to blame for the latter qualities. See also essays by Yang, Wang, Fletcher, and Wills in *The Chinese World Order: Traditional China's Foreign Relations*, ed. John King Fairbank (Cambridge: Harvard University Press, 1968).

4 The end of colonialism and the moral, epistemological, and political ruptures its expiration entailed also produced a seismic shift in humanistic knowledge production, including the emergence of post-structuralism as a forceful critique of Europe's universals as both particularisms and forms of domination. Not only was "Reason" a "white myth," as Derrida put it, but unpacking the myth as such also served as intellectual corollary to political revolution. See Robert Young, *White Mythologies: Writing History and the West* (New York, NY: Róutledge, 1990). (The post-structuralists, it should be said, were not the first to find problems with Enlightenment universalism. A most important earlier critic was Martin Heidegger.) Concurrently with post-structuralism, postcolonialism has developed a sympathetic but distinct line of critique, arguing for a distinct South Asian modernity and attempting to outline what is both "indispensable and inadequate" in European analytics. See Dipesh Chakrabarty, *Provincializing Europe: Postcolonial Thought and Historical Difference* (Princeton, NJ: Princeton University Press, 2000).

5 On the production of contemporary universals see Anna Tsing, *Friction: An Ethnography of Global Connection* (Princeton, NJ: Princeton University Press, 2005). Local co-production of projects previously conceived in Eurocentric terms appears in Kapil Raj, *Relocating Modern Science: Circulation and the Construction of Knowledge in South Asia and Europe, 1650–1900* (Basingstoke: Palgrave Macmillan, 2007); Warwick Anderson, *The Collectors of Lost Souls: Turning Kuru Scientists into Whitemen* (Baltimore, MD: John Hopkins University Press, 2008). Recent studies of the Qing Empire that work from within indigenous political logics include Macabe Keliher, *The Board of Rites and the Making of Qing China* (Oakland: University of California Press, 2019) and Jonathan Schlesinger, *A World Trimmed with Fur: Wild Things, Pristine Places, and the Natural Fringes of Qing Rule* (Stanford, CA: Stanford University Press, 2017).

6 For a critical analysis of such operations see Foucault, *Discipline and Punish.*

7 The Opium War (1839–1842) was settled by the Treaty of Nanjing. The "Arrow" War (1856–1858) led to the Treaty of Tianjin.

8 Andrea Bréard, *Nine Chapters on Mathematical Modernity: Essays on the Global Historical Entanglements of the Science of Numbers in China* (Cham: Springer, 2019), Ch. 7 and 8; Tong Lam, *A Passion for Facts: Social Surveys and the Construction of the Chinese Nation State, 1900–1949* (Berkeley, CA: University of California Press, 2011).

9 Joseph Needham et al., *Science and Civilisation in China*, Vol. 2 (Cambridge: Cambridge University Press, 1954), esp. Ch. 13. Mathematics was also a key part of imperial governance: see Catherine Jami, *The Emperor's New Mathematics: Western Learning and Imperial Authority during the Kangxi Reign (1662–1722)* (Oxford: Oxford University Press, 2012).

10 Literature on the IMC and its origins is extensive. A more recent authoritative survey is Hans J. Van de Ven, *Breaking with the Past: The Maritime Customs Service and the Global Origins of Modernity in China* (New York, NY: Columbia University Press, 2014). On pre-IMC Sino-foreign treaty administration see Fairbank, *Trade and Diplomacy.*

11 In its first decades of operation, Chinese persons staffed only lower-level clerkships and positions that interacted with the Qing Superintendent's office. See Chihyun Chang, *Government, Imperialism and Nationalism in China: The Maritime Customs Service and Its Chinese Staff* (London: Routledge, 2014).

12 Direct interaction between foreigners and agents of Qing revenue collection was a new feature of the treaty regime. Previously, Chinese merchants served as go-betweens. See Paul Van Dyke, *The Canton Trade: Life and Enterprise on the China Coast, 1700–1845* (Hong Kong: Hong Kong University Press, 2007).

13 China. Imperial Maritime Customs, *Provisional Instructions for the Guidance of the In-Door Staff*, First Issue, vol. 5, Service Series, IV (Shanghai: Statistical Department of the Inspectorate General, 1878); S. H. Abbass, *Manual of Customs Practice at Shanghai under the Various Treaties Entered into between China and the Foreign Powers* (Shanghai: Kelly and Walsh, 1894).

14 China. Imperial Maritime Customs, *Inspector General's Circulars*, First Series: 1861–1875, vol. 7, Service Series, IV (Shanghai: Statistical Department of the Inspectorate General, 1879), nos. 4/1861, 1/1862, 9/1863, hereafter cited IG Circular, First Series. Settlements to the Opium War and the Arrow War charged the Qing government with indemnities of 21 million dollars and 16 million taels, respectively.

15 IG Circular, First Series no. 9/1864; China. Imperial Maritime Customs, *Inspector General's Circulars*, Second Series (Nos. 451-600) Vol. 5, 1889–1893, Service Series, IV (Shanghai: Statistical Department of the Inspectorate General, n.d.), no. 476, hereafter cited IG Circular, Second Series.

16 For a brief history of national accounting see Diane Coyle, *GDP: A Brief but Affectionate History* (Princeton, NJ: Princeton University Press, 2014). British statecraft began to consider quantitative information on commodity movements in the seventeenth century.

See, for example, Thomas Mun, *A Discourse of Trade, from England unto the East-Indies* (London: 1621).

17 IG Circular, Second Series No. 524.

18 "Indian and Native Opium in China," *Liverpool Mercury*, December 27, 1888; "News of the Day," *Birmingham Daily Post*, June 11, 1889; "The Foreign Trade of China and Japan," *The Belfast News-Letter*, September 17, 1889.

19 "Address from British Merchants at the Port of Chefoo," Great Britain, *China. No. 5 (1871). Correspondence Respecting the Revision of the Treaty of Tien-Tsin*, vol. 70, Command Papers, C. 389, 1871, 1–2. Hereafter *CRRTT*.

20 Yi Li, *Chinese Bureaucratic Culture and Its Influence on the 19th Century Steamship Operation, 1864–1885: The Bureau for Recruiting Merchants* (Lewiston, NY: Edwin Mellen Press, 2001).

21 Stacie Kent, "Problems of Circulation in the Treaty Port System," in Robert Bickers and Isabella Jackson, eds., *Treaty Ports in Modern China: Law, Land, and Power* (New York, NY: Routledge, 2016), 78–100.

22 "Memorandum on the present condition of the Chinese Empire and its Internal Administration in connection with a Revision of Treaties," In Ian Nish, ed., *British Documents on Foreign Affairs: Reports and Papers from the Foreign Office Confidential Print*, vol. 20: China's Rehabilitation and Treaty Revision, 1866–1869, Series E, Asia, 1860–1914 (University Publications of America, 1994), 64. Hereafter *FO Confidential*.

23 "Memorandum," *FO Confidential*, 64.

24 *Chou ban yi wu shi mo* [Complete record of foreign affairs], Qing nei fu chao ben, Tong-zhi chao, Zhongyang yanjiu yuan lishi yuyan yanjiusuo, Han ji dianzi wenxian, juan 63: 4.1, 5.2, 9.2. Hereafter *CBYWSM*.

25 "Memorial by Tseng Kuo-fan to the Emperor, on the Revision of the Treaty." British Embassy Translation. *FO Confidential*, 159.

26 Li Hongzhang, *Li Hongzhang quan ji* [Complete records of Li Hongzhang], eds. Tinglong Gu and Yi Dai, vol. 3 (Hefei shi: Anhui jiao yu chu ban she, 2008), 3: T6-12-002, 165–167.

27 *CBYWSM*, juan 63: 2.2–3.1.

28 *CBYWSM*, juan 63, 5.2.

29 *CBYWSM*, juan 63, 2.2–3.1.

30 Li, *Li Hongzhang quan ji*, 3: T6-12-002, 166–167.

31 *CBYWSM*, juan 63, 5.2–6.2.

32 Article 28, Treaty of Tianjin.

33 Kent, "Problems of Circulation."

34 Sir R. Alcock to Mr. Hart, 12 July 1868, *FO Confidential*, 280.

35 Sir R. Alcock to Mr. Hart, 12 Jul 1868, *FO Confidential*, 280–283.

36 Sir R. Alcock to Mr. Hart, 21 July 1868, *FO Confidential*, 291.

37 "Reply to the several Propositions, seriatim, of the British Minister as they have been presented and discussed in the Commission, June 28, 1868," in *FO Confidential*, 239–241.

38 Sir R. Alcock to Mr. Hart, 12 July 1868, *FO Confidential*, 282.

39 "Separate memorandum on Revision of the Tariff, for communication to the Tsungli Yamen," 5 September 1868, *FO Confidential*, 244–245.

40 *CBYWSM*, juan 63, 44.1

41 On pearls and fur see Jonathan Schlesinger, *A World Trimmed with Fur: Wild Things, Pristine Places, and the Natural Fringes of Qing Rule* (Stanford, CA: Stanford University Press, 2017); on soybeans see Stacie Kent, *Capital and Commercial Governance in the Late Qing* (PhD dissertation, University of Chicago, 2015), 229–247; on duty exemptions see Yuping Ni, *Customs Duties in the Qing Dynasty, ca. 1644–1911* (Leiden: Brill, 2017).

42 Meng Sen, *Jianyu tongji xu* [A Preface to Prison Statistics, with Tables Appended] (1907) in Bréard, *Mathematical Modernity*, 206.

43 "Memorandum for Sir R. Alcock," *FO Confidential*, 296–298.

44 Sir R. Alcock to Mr. Hart, 10 August 1868, *FO Confidential*, 298.

45 Mr. Hart to Sir R. Alcock, 29 May 1868, *FO Confidential*, 275.

46 Mr. Hart to Sir R. Alcock, 17 July 1868, *FO Confidential*, 289.

47 Sir R. Alcock to Mr. Hart, 10 August 1868, *FO Confidential*, 298–299.

48 Sir R. Alcock to Mr. Hart, 10 Aug 1868, *FO Confidential*, 299; Sir R. Alcock to Mr. Hart, 23 Jul 1868, *FO Confidential*, 294.

49 Mr. Hart to Sir R. Alcock, 7 August 1868, *FO Confidential*, 296.

50 "縱將稅項增加, 亦不能過形蕭索", *CBYWSM*, juan 63, 44.1–44.2.

51 Sir R. Alcock to Mr. Hart, 10 August 1868, *FO Confidential*, 300.

52 The Prince of Kung to Sir R. Alcock, 5 December 1868, *FO Confidential*, 255–256.

53 Scholars have produced substantial literature on the internal contradictions of capital. David Harvey has done much to discuss their spatial implications. See *The Limits to Capital* (London: Verso, 1999 [1982]).

54 Li, *Li Hongzhang quan ji*, 31: G2-04-001, 386.

55 Mr. Hart to Sir R. Alcock, 25 August 1868, *FO Confidential*, 301.

56 Sir R. Alcock to Mr. Hart, 10 August 1868, *FO Confidential*, 299.

57 Mr. Hart to Sir R. Alcock, 25 August 1868, *FO Confidential*, 301.

58 *CRRTT*, 414–421

59 Sir R. Alcock to Lord Stanley, 23 December 1868, *FO Confidential*, 319. In marginalia notes justifying the terms of the agreement, Hart noted that opening the port of Wuhu was of "immense importance, if for the only reason that it makes the Province of Anhui a Treaty-port province for the circulation of British manufactures... If the map must be painted blue there will be something lost, but if painted red the gain will be great." *CRRTT*, 417.

60 *CRRTT*, 428.

61 It seems unlikely simplified tariff would have changed the fate of the agreement. What British commercial interests wanted was more room to grow and the means to do so – more railroads, more coal, more steamship access. The chief complaint was the revision did not do enough to secure these.

62 A discussion of indigo merchants and taxation can be found in Susan Mann, *Local Merchants and the Chinese Bureaucracy, 1750–1950* (Stanford, CA: Stanford University Press, 1987).

63 Nong gong shang bu tongji chu, *Nong gong shang bu tonji biao di yi ci* [First Statistical Tables of the Board of Agriculture, Industry, and Commerce], 6 vols. (Beijing, 1908).

11 Plague and the Global Emergence of Microbiology, 1894–1920

Shiori Nosaka and Matheus Alves Duarte da Silva

Introduction

The emergence of microbiology, particularly its impacts on the field of medicine, is a major subject in the history of science – so much so that it is often framed as one of the major "scientific revolutions." The richness and complexity of the historiography dealing with this subject cannot be captured in a few lines, but we will attempt to highlight a few main points of interest. One of them is the life and career of Louis Pasteur himself, his first experimentations with anthrax and rabies vaccines, the creation of a scientific school at his laboratory in Paris, and the resulting transformations in French agriculture, industry, and society.[1] Other historians have studied Pasteur's German counterpart, Robert Koch. They have described the work that led Koch to identify tuberculosis and cholera bacilli, his opening of a school in Berlin, and the rivalries between him and Pasteur, which mirrored pre-World War I tensions in Europe.[2] Most recently, historical studies have focused on another episode of the production of microbiological knowledge in Paris and Berlin where, in a more collaborative effort, researchers created the antidiphtheric serum, which was used on human populations on a large scale for the first time in 1894.[3] Alongside these European accounts, other authors have examined the emergence of microbiology in other regions of the world starting in the 1890s. Some have focused on the exportation of the new scientific discipline to European colonies, looking mostly at British India and parts of the French Empire. Others have studied how microbiology was adapted in independent states like Brazil or Japan. In their accounts, historians have sometimes brought to light the violence associated with this process and the power imbalance that these scientific exchanges fostered. However, these exchanges went beyond violence and domination, and some historians highlighted the use of microbiological tools by colonial elites or independent states to solve local/national problems.[4]

Despite the richness of these works, taken together, they drive the idea that the core of microbiological knowledge and expertise in medicine at that time – research and production of sera and vaccines – was almost completely created in France and Germany before being exported almost as a set of stable techniques (as a "black box," in the words of Bruno Latour) to be applied in the rest of the world. The present essay aims to problematize this general assumption and to provide some

DOI: 10.4324/9780367814540-13

insights about the possibilities of a global history of microbiology. Indeed, in the years spanning the creation and stabilization of the antidiphtheric serum in 1894 and the onset of World War I, microbiology became a global science for the first time. We aim to show that this process was not only the consequence of French and German increasingly widespread networks or because some countries started to establish their own national traditions in the field of microbiology, as emphasized by many historians. Although this certainly was happening, we argue that micro-biology became a global science at this moment because scientists working in the Americas, Europe, and Asia started to investigate the same disease, plague, allow-ing them to compare their experiences and to share objects and medical theories.

To show the potentialities of a global history of microbiology, this chapter will focus on the Pasteur Institute of Paris and on its connections with researchers based in India, Japan, and Brazil, three places which deeply distinct political situations, brought together both by the plague pandemic and by microbiology. In this histori-cal account, we will follow a circulatory approach, as discussed by Kapil Raj. For Raj, the circulation of knowledge should not be reduced to *mobilities* but should be understood as a dynamic process of transnational and transcultural encounters, exchanges, resistance, and negotiations that produce new knowledge.[5] To develop such an approach to our case study, we will stress, on the first part of the essay, how the Pasteur Institute of Paris played an important role in the discovery of plague etiology, in 1894, and in the production of the first antiplague serum, which was tested in India in 1897 with unforeseen international outcomes. In the second and third parts, we will examine how microbiological research on antiplague sero-therapy fast expanded in Japan and Brazil. We will discuss how these countries faced the disease by adapting or creating new sera, and how Japanese and Brazilian scientists sought to reinforce their local and international positions thanks to these objects. The fourth part will return to the beginning of the pandemic, showing the emergence in India and France of three techniques of antiplague immunization: vaccination, vaccination by the serum and mixed vaccination (combining the vac-cine and the serum). The fifth and the last parts will discuss how Brazil and Japan interacted with, challenged, and changed these techniques, producing new knowl-edge of antiplague immunization capable of reshaping the research conducted at the Pasteur Institute of Paris.

Doubts and Hopes about Antiplague Serotherapy (1894–1899)

The antidiphtheritic serum was the first successful therapeutic serum created by microbiologists and arguably the first microbiological development capable of pro-viding treatment on a large scale. Between 1889 and 1890, scholars at the Pasteur Institute in Paris and at the Charité Hospital in Berlin noticed that some bacteria, such as those causing tetanus and diphtheria, had the capacity to secrete a toxin responsible for the infection and death of human patients. They further noticed that the blood of rabbits could neutralize the toxin when it was present in small doses. This neutralization, barely understood by then, was apparently due to antitoxic sub-stances (antibodies, in modern medical terms) produced by the animal's organism.

These first attempts turned into bigger serological research: larger animals such as horses were injected with the diphtheritic toxin for a few weeks and then bled. Their blood serum, containing antibodies, was used to treat other animals and then humans, primarily children infected with diphtheria. The first trials in Paris and in Berlin in 1894 were seen as a success, as mortality sank among patients who had been administered the serum. Henceforth, a hope emerged among microbiologists and the public in general that this first achievement could be translated into a cure for other infectious diseases, and that new sera could be produced to treat them.[6] The plague pandemic, which broke out in 1894, was the first opportunity to test this optimism.

During the second half of the nineteenth century, at least in Western Europe, plague was considered a disease of the past. Cholera had taken its place as the deadliest global scourge. Indeed, nine international conferences were held between 1851 and 1895 to organize responses against cholera and scientific missions were sent to places that were thought of as possible "homes" of cholera, such as Egypt and India.[7] The situation started to change in 1894 when an epidemic of bubonic plague broke out in the British colony of Hong Kong. In less than five years, the disease took root in many parts of Asia, Africa, and the Americas and reached a few Western European ports on some occasions. To face this pandemic, which caused millions of deaths, especially in India, international sanitary conferences were organized in 1897, 1903, and 1910–1911. Moreover, scientific missions were dispatched to study plague in different places, particularly in the first years of the pandemic.[8]

Three of these scientific missions were led by Alexandre Yersin, former *préparateur* at the Pasteur Institute in Paris and colonial doctor in Indochina since 1893. Yersin's first mission took place in Hong Kong and started just after the beginning of the outbreak in 1894. In that British colony and in competition with the Japanese doctor Shibasaburo Kitasato, Yersin described for the first time the plague bacillus and sent some samples to the Pasteur Institute in Paris. A few months later, Yersin returned to France and, helped by his colleagues Albert Calmette and Amédée Borrel, developed in Paris a first prototype of an antiplague serum, which started soon to be produced in France and in Nha Trang, Indochina, in a laboratory directed by Yersin. The second mission of Yersin took place in 1896, when he traveled to the Chinese cities of Canton and Amoy (present-day Guangzhou and Xiamen, respectively) to test for the first time the antiplague serum in humans. Using up all of his very limited stock of serum, Yersin saved 24 of 26 patients and considered the drug a success.[9] In the end of 1896, plague cases were reported in Bombay (present-day Mumbai), the economic hub of British India. Thanks to the previous success of Yersin in China, the Municipality of Bombay invited him to apply the antiplague serum there.[10] The third mission of Yersin, this time in India, started in March 1897. This mission was carried on and concluded by his colleague Paul-Louis Simond, who substituted Yersin in June 1897 and remained in India for more than a year.[11]

Yersin and Simond carried out several tests in India with slightly different antiplague sera manufactured by the Pasteur Institutes of Paris and Nha Trang.

Contrarily to the initial experiments in China, the tests in India suggested a more negative scenario, where the various antiplague sera possessed an efficacy of only 40%, which could be due to intrinsic problems of the product or the virulence of the Indian outbreaks. In addition to these poor results, the years 1897–1898 witnessed a fierce competition for antiplague serotherapy in Bombay. Indeed, the plague in India had attracted other European scientific missions from countries such as Russia, Germany, Austria, or Italy. Hailing from the latter, for instance, were the doctors Alessandro Lustig and Gino Galeotti from Florence University, who had invented in early 1897 a different antiplague serum.[12] The competition between the French and Italian sera started at the end of 1897 and was partially resolved in 1898 when the Municipality of Bombay decided to sponsor a local production of Lustig's drug in the new Municipal Laboratory.[13]

The poor results observed in Bombay as well as the scenario of global competition pushed Émile Roux, then vice-director of the Pasteur Institute in Paris, to recalibrate the French antiplague serum after 1898. Roux abandoned the previous technique – the inoculation of exclusively dead bacilli – and decided to inoculate the horses with dead plague bacilli followed by injections of virulent plague bacilli, which he believed could produce stronger results. This new antiplague serum was first tested by Albert Calmette and Alexandre Salimbeni in Porto, Portugal, during a plague outbreak starting in August 1899. The treatment with it resulted in an 85% recovery rate among 142 patients, which allowed the Pasteur Institute scholars to claim that the problems observed in Bombay with their serum had been solved.[14] However, Lustig and Galeotti disputed this assertion. They argued that plague had lost part of its virulence when it moved into Europe. For the Italians, the apparent proof of the efficacy of the new French antiplague serum was nothing but a misunderstanding, and the results obtained by Calmette and Salimbeni only showed that the plague of Porto was of a milder type than that of Bombay.[15] Precisely during this period of doubt and hopes about the capacity of microbiology to halt the epidemic and cure patients, the disease broke out in Japan and Brazil, triggering similar as well as new questions about the efficacy of antiplague sera.

A National and Imperial Question: Antiplague Serotherapy in Japan (1899–1911)

The plague first reached the Japanese territory in 1896, touching Taiwan, which had been a Japanese colony since 1895. A scientific mission arrived in Taipei in May 1898, led by Kiyoshi Shiga, a researcher at the Institute for Infectious Diseases in Tokyo. Shiga tested three antiplague sera prepared with a strain of bacteria stored in Tokyo, called "Kitasato bacilli," isolated in Hong Kong in 1894. The pursuit of serum research with Kitasato bacilli was highly significant for Japanese researchers: if they could prove the effectiveness of this serum, they would confirm the validity of Kitasato's discovery of the plague bacillus, which Kitasato disputed with Yersin. However, the clinical results were not promising: the mortality rate after serum injections was 33.3%, whereas the surgical removal of the buboes entailed a mortality of 21.75%.[16] The following year, when the first cases

were detected in mainland Japan, the situation became untenable for those who believed Kitasato's bacilli were the causative agent of the infection, because Kitasato
himself recanted.[17] Simultaneously, as soon as the presence of plague in Kobe was
announced, the French consul in Japan wrote to Paul Doumer, Governor-General
of Indochina, requesting the serum prepared by Yersin in Nha Trang.[18] Doumer
provided the antiplague serum to the Japanese government, requesting no payment
for the products or shipping.[19] Serotherapy in the mainland was therefore undertaken with the French product, which was used in a hospital in Osaka to treat 12
patients. But the therapeutic benefits of the serum were unconvincing: only one
patient survived, and this patient was also treated with purulent drainage. Japanese
doctors concluded that although Yersin had identified the plague bacillus successfully, his serum needed to be improved.[20]

Following these clinical studies, the production of antiplague serum was taken
up by the Serum Institute in Tokyo, a microbiological laboratory dedicated to the
fabrication of antidiphtheric serum and rabies vaccine. The laboratory borrowed
a "safe" manufacturing method suggested by the German plague commission in
India: injection of heated bacilli in horses, to avoid the risk of spreading living
bacilli in the surrounding area of the laboratory.[21] Sahachirô Hata, from the Institute for Infectious Diseases in Tokyo, in charge of antiplague sera production,
considered that the new serum manufactured in Japan was more efficient than Yersin's, judging from both serological and animal tests. The new Japanese antiplague
serum was used experimentally in Taiwan from March to July 1900. Ki'ichi Tsukiyama, who administered this serum in the city of Tainan, concluded that the drug
could indeed reduce bubonic plague mortality. Among 54 patients treated with the
product, 66% recovered, compared to 37.5% in a control group.[22]

However, Hata did not seem to be satisfied with these results. He claimed that
the clinical trial did not show the same results as animal experimentation. From
clinical experiments in Osaka in the last months of 1900, he noticed that, even
with the improved serum, several high-dose administrations had to occur to keep
patients alive.[23] With these clinical difficulties, the capacity to manufacture serum
and the selection of a safe transfer medium became serious problems for serotherapy. Even though the epidemics was not catastrophic in the mainland – between
November 1899 and August 1914, 1,158 cases were identified, and 934 patients
died – plague patients in Taiwan often numbered over 1,000 per year between
1896 and 1914; that number rose to 4,500 cases in 1904.[24] This situation led Japanese scholars to focus more on the amount of serum that had to be administered
than the quality of the serum itself. In 1908, Gorôsaku Shibayama pointed out
that results of serotherapy carried out in Japan were consistently inferior to results
elsewhere because they were not able to produce and administer high doses of
serum.[25]

The practical issues raised by plague serotherapy were also highlighted internationally, mainly after an outbreak of pneumonic plague in Northern China, particularly in Manchuria, in 1910 and 1911. During the epidemic, 2,185 phials of serum
were sent from Tokyo to local Chinese authorities in Manchuria, but Japanese
microbiologists could not follow how these products were used there.[26] Clinical

studies for serotherapy during the epidemic were carried out in Harbin on 42 patients with a Russian serum produced in Kronstadt, as well as the French serum sent from Paris. Paul B. Haffkine, the Russian doctor in charge of this treatment, reported that no one survived.[27] In Dairen, a Japanese doctor treated 13 patients with the serum from the Institute for Infectious Diseases in Tokyo, which led to four recoveries.[28] However, at the international plague conference held by the imperial Chinese government in Mukden in April 1911, participants concluded that it was doubtful whether the Dairen patients were all cases of pneumonic plague; even if serotherapy was useful under certain conditions – several injections of a high dose of serum as early as possible –this therapeutic method was considered impractical in epidemic situations of pneumonic plague. A brief report on the conference, distributed by the Chinese minister of Foreign Affairs one month later, summed up the arguments: "[t]he chief factor in the decline of the epidemic has probably been the preventive measures which were enforced in accordance with scientific methods or by efforts of the people to protect themselves," and concluded that therapeutic methods were inefficient.[29]

Although antiplague serotherapy spurred Japanese laboratories to produce a large amount of serum, to such an extent that they could send stocks to emergency epidemic zones outside Japan, critical observations on the large quantity of serum required for therapeutic use were repeated over the years. The gap between the successful production of sera in the laboratory and the practical limitations of serotherapy led microbiologists to acknowledge the problem and fostered their interaction on an international level. Even though the efficacy of serotherapy was not demonstrated in this case, Japanese microbiologists' attempt to innovate the practice of serotherapy is to be underlined here: it was this attempt, in the Japanese case, that allowed global exchanges around antiplague serotherapy.

Reinforcing Local and Global Positions in Brazil (1899–1910)

Until the end of the nineteenth century, yellow fever was the most important subject of microbiological investigation in Brazil. The disease prompted a few attempts to produce new sera and vaccines in the country and led to the establishment of laboratories in its capital, Rio de Janeiro.[30] However, these measures against yellow fever were not conclusive, at least not until 1903, when a successful campaign to eradicate the mosquitoes responsible for transmitting the disease started in Rio de Janeiro. Thus, when the twentieth century began, microbiology held good promise in Brazil, but it had a limited impact on Brazilian public health.

This panorama started to change when the plague arrived in Brazil. The first cases were identified in the port of Santos in October 1899; from there, it spread to many other cities like São Paulo and especially Rio de Janeiro, where it became a major sanitary crisis during the first decade of the twentieth century.[31] At the beginning of the outbreak, microbiologists in Brazil faced the limitations of antiplague serotherapy, discussing, for instance, the results obtained in India and whether the new technique was able to curb the outbreak in Brazil.[32] At that moment, Brazil, like Japan, imported sera from Paris but also from Messina, Italy.[33] From 1901 on, two local laboratories established in 1899 and 1900 – Manguinhos, in Rio de Janeiro, and Butantan, in São Paulo – began to manufacture sera and substituted the imported drugs. When their operations started, both Brazilian laboratories decided to apply the French technique tested in the city of Porto to produce their own sera: an injection of dead bacilli followed by an injection of virulent bacilli.[34] In 1902, Butantan streamlined the process by sidestepping the injection of dead bacilli, and the idea was emulated by Manguinhos in 1905.[35]

Despite initial hesitations, the antiplague sera produced by Manguinhos and Butantan succeeded in saving most of the patients in São Paulo and Rio de Janeiro during the first decade of the twentieth century. In the small outbreaks of São Paulo, the death rate always remained below 40% among treated patients; in Rio de Janeiro, fatality rates among patients treated with the Manguinhos serum decreased over the years (from 35% in 1901 to only 7% in 1907), which according to some Manguinhos researchers was proof that the serum had continuously improved in quality.[36] Although scholars working in both laboratories saw these results as remarkable, they might have been due to the attenuation of the plague bacilli, which could happen naturally over the course of an epidemic. Indeed, the disease was never as virulent in Brazil as it was in India, and the number of annual cases (500 on average in Rio de Janeiro, the center of the epidemic in the first decade of the twentieth century) always matched the capacities of the Brazilian laboratories to produce sera.[37] Nevertheless, Brazilian scholars insisted that the good results of the sera produced in Manguinhos and Butantan were due to their own innovations.

Following this statement, the directors of Butantan and Manguinhos (Vital Brazil and Oswaldo Cruz, respectively) tried to show to local and international audiences that their drugs were superior to those produced in India and in Europe. They used two strategies to accomplish this. First, they sent their sera abroad to be tested. Cruz forwarded some flasks of the Manguinhos serum to Wilhelm Kolle in Berlin, who injected it in animals and compared it to similar drugs. The positive results in Berlin were used by Manguinhos researchers to make a case for the efficacy of their product and the importance of the laboratory to the Brazilian government.[38] Vital Brazil was more audacious and sent the Butantan serum to Bombay, the global epicenter of the pandemic. Throughout 1902 and 1903, the Butantan lab sent almost 40% of its production – more than 600 tubes – of the antiplague serum manufactured in São Paulo to Bombay.[39]

In Bombay, Butantan's serum was tested on humans, and the results were compared with sera produced locally, such as the Lustig one, and abroad, such as the Pasteur Institute's serum. William Bannerman, the director of the Plague Research Laboratory, discussed the results of these tests in 1905. He argued that all these products, including the Butantan serum, were useless. Therefore, he declared that the mode of fabrication of these products must change, and that their antitoxic potential should be increased.[40] This statement impacted plague serotherapy research in Bombay and all around the world. Some laboratories abandoned it altogether,[41] while others aimed to develop a new serum using only toxins extracted from the plague bacillus. That was the case with Butantan. Indeed, right after receiving the results from India, Vital Brazil and his colleagues tried to create an antitoxic antiplague serum (the serum sent to Bombay was believed to act only against the plague bacilli but not against its toxins). In 1907, they announced promising results for this new serum after tests in São Paulo. However, this new drug was never sent back to Bombay.[42]

Albeit this material circulation of sera played an important role in convincing the rest of the world that Brazilian laboratories could produce good serum, a second strategy was also pursued by some Brazilian doctors. Indeed, to confirm the efficacy (and sometimes superiority) of their antiplague sera they decided to produce yearly statistics concerning plague serotherapy. This strategy was mainly adopted by Manguinhos, due mostly to logistical reasons: the city of Rio de Janeiro saw repeated outbreaks of the bubonic plague from 1900 on, and it was possible for the actors working there to accumulate data for many years. This was more difficult to pursue in São Paulo, where the outbreak was intermittent. The idea behind the gathering of these data was first to prove to the Brazilian government that plague mortality was decreasing thanks to the effectiveness of the sera produced by Manguinhos, and second to secure funds for the continuation of the laboratory's operations.[43] In addition, these statistics could also strengthen the position of Manguinhos beyond Brazilian borders. To fulfill this project, Brazilian doctors relied on their French colleagues, helped by the fact that from 1901 to 1905, a scientific mission to study yellow fever led by three members of the Pasteur Institute – Émile Marchoux, Alexandre Salimbeni, and Paul-Louis Simond – stayed in Rio de Janeiro. Because the French laboratory continued to keep an interest in plague serotherapy, the statistics concerning plague treatment in the Brazilian capital circulated in France, where they were used to prove the efficacy of the Manguinhos serum and antiplague serotherapy in general, including the Pasteur Institute's own antiplague serum. In that sense, Brazilian data were used in France precisely to counterbalance doubts raised about the efficacy of serotherapy after the publication, in 1905, of the tests conducted in India with many sera.[44] In short, the data produced in Brazil by Manguinhos, circulating between Brazil and France, fulfilled multiple and distinct purposes of several actors, which can be better understood when the positive views of serotherapy that arose in Brazilian and French exchanges are analyzed side by side with the critical views coming from India.

To conclude this section, it is worth highlighting two points. First, antiplague sera produced lukewarm results on a global scale: on occasion and in some places, as in the Brazilian cities, they saved patients, although frequently in a lower rate than the antidiphtheritic serum. In other epidemic situations, however, they were almost unable to prevent deaths caused by plague, as was evidently the case in India. Second, research on antiplague sera was conducted mostly outside Europe – in India, Japan, and Brazil – but it influenced some scholars working there, as shown by the publication in Paris of the data produced in Brazil by the Manguinhos laboratory. In summary, studies on antiplague sera in India, Brazil, and Japan gave birth to a new and global vision of serotherapy, which acknowledged its potentialities but also its limits and inherent instability.

The Emergence of New Techniques of Antiplague Immunization (1896–1899)

The plague pandemic was also an opportunity to (re)invent immunization practices, a process that fostered new relationships between scientists from various countries and triggered a new wave of knowledge production and circulation. As with the debates on treatment, India played a pivotal role in discussions around antiplague immunization. When the bubonic plague reached Bombay in September 1896, the Government of India sent one of its most accomplished microbiologists, Russian-born Waldemar Haffkine (uncle to the previously mentioned Paul B. Haffkine), to manage the crisis. He had been working in India since 1893, testing a vaccine against cholera that he invented at the Pasteur Institute of Paris. In Bombay, Haffkine attempted to use his experiments with the cholera vaccine in India to create a new prophylactic. He decided to use not only the dead remains of bacilli, as he had in his previous vaccine, but also toxins derived from the plague bacillus.[45] Haffkine started administering this antiplague vaccine in early 1897, manufacturing it in some improvised laboratories until 1899, when the more suitable facilities of the Plague Research Laboratory were inaugurated. Haffkine's vaccine was then massively distributed in Bombay and everywhere in India, and its production reached 500,000 doses a year. In 1902, an accident in the village of Malkowal (or Mulkowal), in the Punjab region, plunged down the production of antiplague vaccine for a while. However, Haffkine's vaccine quickly recovered its credibility and, at least until the end of World War I, remained the main microbiological public health solution against the plague in India.[46]

During these years of intense production and application of the vaccine, Haffkine and his colleagues offered some observations about the effects of the prophylactic. By comparing inoculated and control groups, they concluded that the vaccine reduced the incidence of the disease among the inoculated population by almost 80%. They also agreed that it was able to protect the inoculated for the duration of the plague season (six months on average) but probably not for the duration of their lifetime. Despite its limitations, the vaccine's outcomes were positive enough to make Haffkine and his colleagues encourage its administration where possible in India.[47]

Yersin and Simond arrived in Bombay during the first years of Haffkine's vaccination campaign. As we pointed out before, their aim in India was to study the plague and determine the curative powers of the Pasteur Institute's antiplague serum. However, Yersin and Simond also believed this serum could act as a vaccine and aimed to test this hypothesis in India. The idea behind serum vaccination was that, since serum was nothing but antitoxic and antimicrobial material, giving it to a healthy person would create a temporary barrier against future infection, which would function as a form of "passive immunization." Following this statement, Yersin and Simond vaccinated with serum more than 1,000 people in Bombay and surrounding areas between 1897 and 1898, concluding that the serum produced an almost instantaneous and harmless immunization that lasted around two weeks. With further observations, they also concluded that Haffkine antiplague vaccine could be dangerous: its mode of production was not safe because of the presence of toxins in the vaccine, and immunizing effects only started a few days after inoculation, leaving the vaccinated person vulnerable for a while.[48] Thus, Simond and Yersin recommended the use of the antiplague serum to the French authorities not only as treatment but also as a vaccine. Following this advice, vaccination by serum was adopted from 1898 on in some French lazarettos and in Indochinese ports.[49]

When the plague reached Porto in late 1899, a new immunization practice was proposed against the disease. Following their colleagues' model, Calmette and Salimbeni vaccinated 600 people with serum in this Portuguese city. Although they observed that the process was safe, they concluded that a two-week immunization was not sufficient to prevent people from getting the plague during an epidemic season. They also tested Haffkine's vaccine on animals and observed again that it could be harmful both because the vaccine contained plague toxins and because of the delay between vaccine administration and the onset of immunity. However, active vaccination seemed to offer protection for longer periods than the serum. As both drugs presented pros and cons, Calmette and Salimbeni proposed to add some drops of the antiplague serum to the vaccine and inject them together. Their idea behind this mixed vaccination scheme was to combine the long-term active immunization provided by the vaccine and the short-term, but immediate, passive immunization elicited by the serum. Despite the compelling reasoning, mixed vaccination remained merely a theoretical possibility, since Calmette and Salimbeni did not use it on humans.[50]

Mixing Immunizations in Brazil (1899–1910)

Mixed vaccination would only come into practice in the new centers of microbiological knowledge production that were emerging at the time, Brazil above all. Indeed, Rio de Janeiro was one of the first places in the world to combine the three practices – active vaccination, serum vaccination, and mixed vaccination – into one sanitary strategy. Plague vaccination started in the city following the arrival of an Italian doctor, Camillo Terni, in the first days of 1900. Terni had modified Haffkine's vaccine in his laboratory in Messina, and instead of cultivating the bacilli in broth, as Haffkine, Terni preferred to use the peritoneum of guinea pigs as a medium because it speeded the production of plague bacilli. Terni came to Brazil in an official mission to test his new vaccine there. After some months of trial, Brazilian authorities approved Terni's vaccine and gave him access to the facilities of the *Laboratório Bacteriológico Federal*. Until he returned to Italy in late 1900, Terni used the facilities to produce his vaccine, and inoculated almost 40,000 people, roughly 10% of Rio de Janeiro's population.[51]

Despite the relative success of Terni's vaccine, Oswaldo Cruz, the director of Manguinhos, initially judged not only the product but antiplague vaccination in general a dangerous public health procedure. Cruz had some reasons to be skeptical. First, he had tested Terni's vaccine on several guinea pigs and found that the animals died after receiving it. Second, in part because of his training in Paris and his familiarity with Calmette and Salimbeni's work, Cruz believed that antiplague

vaccines could be harmful during an epidemic in human populations, mostly because of the toxins present in the liquid. Therefore, Cruz argued against plans to
conduct massive vaccination with Terni's vaccine in 1900. Nevertheless, he was
contested by Terni and some Brazilian doctors, who argued that these risks had
not been observed in India, by Haffkine, or in Rio de Janeiro, by themselves.[52] In
1901, Cruz's opinions partly changed and Manguinhos started delivering a new
antiplague vaccine, cultivated in agar-agar and using a more rigorously standardized process.[53]

Just after the delivery of the first lot of antiplague vaccine produced in Manguinhos, Cruz proposed a new immunization strategy against the plague, communicated in a public conference in May 1901. He divided his propositions into three
main tiers. First, he suggested vaccinating people with the Manguinhos antiplague
vaccine outside periods of epidemic. Second, because of the risks intrinsic to the
antiplague vaccine, he recommended using the mixed vaccination during the epidemic. Finally, concerning people living in infected households or who had been
in contact with an infected patient, Cruz stated that they should only be vaccinated
with the Manguinhos serum. In sum, Cruz sought to create a new public health
strategy, a sort of synthesis, that brought together practices developed some years
before in India and Europe.[54]

In 1903, Cruz became the Director of the Federal Department of Public Health
(*Diretoria Geral de Saúde Publica*, thereafter DGSP), whose jurisdiction was limited to the city of Rio de Janeiro. Right after taking charge, he decided to target
three epidemic diseases: yellow fever, plague, and smallpox. Against smallpox,
Cruz proposed that Jennerian vaccination should become mandatory. The "vaccine law," approved in November 1904, started a one-week insurrection, one of
the largest urban unrests in Brazilian history. The population of the city destroyed
public infrastructure, attacked the police, and threatened the President, Rodrigues
Alves, who was forced to flee the city. After violent battles between the people and
the police, and the bombing of Rio de Janeiro by the Navy, the federal government
crushed the insurrection. Despite the victory, the law was discarded, and the idea of
imposing compulsory vaccination against smallpox was abandoned.[55]

This rebellion had a collateral impact in the campaign against the plague, since
the DGSP was not able to enforce by law the immunization solutions proposed
by Cruz in 1901. It became a matter of personal decision whether to pay for serum and the vaccine in a pharmacy, go to a public hospital to receive the drugs
free of charge, or neither. Although there are documents showing the concomitant
application of the three techniques – vaccination, serum vaccination, and mixed
vaccination – in Rio de Janeiro in the first decade of the twentieth century,[56] Cruz
acknowledged in some official reports that the Brazilian capital's population, wary
of vaccines in general, did not seek the antiplague vaccine in the scale he considered ideal. There was an exception, however, to this defeat of "biopower:" the
report of someone infected with the plague. According to Brazilian law, all people
living in the same house as an infected patient were to be mandatorily isolated.
Nevertheless, if they accepted vaccination, they regained freedom of movement. In
this specific situation, mixed vaccination was used.[57] After 1910, bubonic plague

became under control in Rio de Janeiro. Thus, the immunization strategy – based on a mix of different approaches – began to disappear progressively from the city and the DGSP decided to focus mostly on urban intervention, such as rat extermination in the last infected spots.[58]

Plague Immunization in Japan: between the Global and the Local (1899–1920)

The three practices of immunization mentioned above – active immunization with vaccine, passive immunization with serum, and mixed vaccination – were also tested in Japan during epidemics in Kobe and Osaka, starting in November 1899. The main purpose of the trials was to determine how to avoid the potential risk that antiplague vaccines could pose in epidemic situations, a question pointed out by Yersin and Simond in Bombay, by Calmette and Salimbeni in Porto, and by Cruz in Rio de Janeiro. The director of the Momoyama Hospital in Osaka, Masahiro Iwai, administered all three types of immunization on infected patients' family members and medical staff of the hospital, at their request. To immunize six people who had had contact with patients, Iwai only used the antiplague serum that had been sent from Nha Trang by the French colonial governor. With 14 people who had not had contact with plague patients, Iwai used Haffkine's vaccine, which a Japanese trade company had imported from India.[59] Among the six contact cases who were vaccinated with the serum, one died from the plague, whereas two of the 14 people who received Haffkine's vaccine-only product died. Thus, Japanese microbiologists established a hypothesis that these three victims were already infected when they got vaccinated, a hypothesis which led these scientists to consider that the immunizing products themselves were harmless but could be fatal for humans who had already been infected. They concluded that both vaccination strategies – with a vaccine or a serum – would be less effective in an epidemic situation, where many people could be already infected.[60] Meanwhile, four medical staff working with plague patients received the mixed vaccination (vaccine and serum). The products were prepared by Iwai himself, following techniques developed by Kiyoshi Shiga, a researcher at the Institute for Infectious Diseases in Tokyo. Since none of the injected staff died from the plague, Iwai concluded that mixed vaccination was a better method than the others.[61]

Shiga's techniques of mixed vaccination were not only a result of the outbreak of bubonic plague, but mainly of his previous research on dysentery, a virulent endemic disease in Japanese rural areas. Shiga identified the dysentery pathogenic agent in 1897, beginning shortly thereafter the production of antidysentery vaccines.[62] Shiga was concerned about the strong reactions caused by vaccine injections in Haffkine's vaccination programs against cholera and plague in India, so he adopted the German method of cultivating in agar-agar medium to prepare his antidysentery vaccine. Despite this, he continued to observe strong reactions following injection. This led him to modify two processes: he decided to crush the dysentery bacilli's bodies in an agate mortar, and then to mix them with antidysentery serum. He presented this mixed preparation in the *Journal of Microbiology*

[*Saikingaku zasshi*] in September 1899, shortly before Calmette and Salimbeni published their paper in the *Annales de l'Institut Pasteur* in December 1899. In his paper, Shiga mentioned that thousands of people vaccinated with his preparation in Nagano, a city in the mid-northern region of Japan, showed scarcely any reactions.[63] To Shiga, the absence of undesirable reactions meant that the vaccine was properly absorbed in the body. In this sense, Shiga's mixed vaccination against dysentery was the fruit of clinical observations, rather than theoretical deduction or animal experimentation, as was the case of Calmette and Salimbeni's proposal of the antiplague mixed vaccination scheme.[64]

To fight the plague in Kobe, Shiga applied a similar method of mixed vaccination as originally used against dysentery: after preparing the culture of plague bacilli in an agar-agar medium, he crushed them and heated them up to kill the bacilli. The inoculation happened in two steps: first, patients received a mixed vaccine (equal proportions of Shiga's plague vaccine and antiplague serum); second, they received a shot of the pure vaccine two or three days after the mixed shot. Shiga administered the two-step immunization to 47 medical staff in Kobe and Osaka. When Shiga published his work on plague vaccination in January 1900, he mentioned Calmette and Salimbeni's proposal. However, he stressed the difference between the two approaches: while the Pasteur Institute researchers had imagined mixed injection of serum and vaccine as a way to combine active and passive immunity, Shiga's focus was to facilitate the development of immunity.[65] He aimed, above all, to curb the vaccine's side effects, which were worrying with Haffkine's product. After adopting this relatively safe immunization method, the Tokyo Serum Institute began in May 1900 mass production of antiplague vaccines cultivated in agar-agar culture.

Even though the first initiative of antiplague immunization was Shiga's two-step process that started with a mixed-vaccine shot, Japanese doctors ended up adopting a different approach when massive vaccination campaigns began in Osaka in July 1900. Contrary to Shiga, his colleagues Shibayama and Hata considered that the mixture of serum and vaccine would not accelerate the absorption of the latter in any way. They agreed with Calmette and Salimbeni's idea that a combination of serum and vaccine could confer two types of immunity onto subjects living through epidemic situations. But since the active immunization conferred by vaccines could be a fatal danger in case the vaccinated subject had already contracted the plague, Shibayama decided to inject the two products separately. First, the antiplague serum was to be injected into subjects who might have contracted the plague. Then, if they did not show any sign of infection several days later, they would receive a shot of active vaccine. This split use of serum and vaccine advocated by Shibayama was limited to medical workers, while the inoculation of the bacilli cultivated in agar-agar and then killed became the main method of optional prevention for large populations. Between July 1900 and March 1901, 61,603 inhabitants were vaccinated in Osaka, 21,361 in Kobe, and 4,456 in Wakayama.[66]

Vaccination also started in Taiwan in September 1900, when one customs officer died from the bubonic plague in the city of Tainan. Tsukiyama, a hospital doctor in Tainan, injected Haffkine's vaccine into 135 workers of the customs office, and 75 medical staff in the hospital as a preventive measure.[67] Tokyo vaccines, produced in agar-agar culture, began to be shipped to Taiwan that same month, but it seems that Tsukiyama continued to use Haffkine's vaccine as the main prophylactic measure.[68] In November 1900, Tsukiyama wrote to an officer of the colonial government, emphasizing that since Haffkine's vaccine had garnered the people's trust, the situation was now favorable to expanding vaccination efforts.[69] However, when Tsukiyama demonstrated in an article published in 1901 that 25,321 Taiwanese and Japanese people had been inoculated between August 1900 and June 1901, he mentioned the Tokyo vaccine as the only product administered in his vaccination program.[70] Indeed, the vaccine supply from Tokyo started to increase at the beginning of 1901, but the absence of any statistical comparison with Haffkine's product may be due to Tsukiyama's intention to demonstrate Taiwan's example as a success of the Japanese technique. To make convincing this triumph, he emphasized a decreased morbidity of 30.6% among the vaccinated. This result allowed the generalization of antiplague vaccination in Japanese imperial territories during the first decade of the twentieth century.

In short, three types of immunizing practices were examined and mobilized in Japan in its local epidemic situation. With these practices, government officials and scientists sought to use these methods according to different population groups (medical workers, family members of positive cases, and ordinary citizens) or epidemiological situations (highly epidemic or not). In that scenario, mixed vaccination strategies were developed to respond to local needs – building immunity rapidly for a long time. Therefore, it is worth stressing that while the mixed vaccination method derived from Calmette and Salimbeni's hypotheses, the effective implementation of this new technique of immunization took place in Brazil and Japan, being developed and tested locally. Moreover, the application of antiplague vaccines in large scale, without the concurrence of sera, likewise happened mostly in Indian, Brazilian, and Japanese frameworks. Because of that situation, the Pasteur Institute started to seriously consider the data that was emerging from these places.

In fact, in the first decades of the twentieth century, the immunization strategy supported by the Pasteur Institute against the bubonic plague in France and in French imperial territories – based on serum vaccination and mixed vaccination – changed over time. Researchers almost completely abandoned both solutions and came to support the active immunization induced by the vaccine. The process leading to this transformation started in the Pasteur Institute facilities in 1902, when Alexandre Besredka set out to develop a new antiplague vaccine. To reduce the toxicity of Haffkine's technique, Besredka chose to put the killed plague microbes in contact with an emulsion of antiplague serum, which was then discarded after a few hours. Thus, Besredka's vaccine was atoxic, unlike his colleagues Calmette and Salimbeni's mixture, which carried toxin-laced serum. Indeed, he believed the latter was useless, since some properties present in the serum could cancel the prophylactic properties of the vaccine when injected together.[71] From 1902 on, other criticisms against mixed vaccination arose in France, especially on the assumption that the application of small doses of sera, as it was the case in the mixed vaccination, could provoke fatal anaphylactic accidents.[72]

Even though Besredka's vaccine was believed to be safer and more efficient than Haffkine's, it was not produced *en masse* at the Pasteur Institute, perhaps due to manufacturing costs. Instead, in 1909, the French laboratory started delivering another kind of antiplague vaccine, where the plague bacilli were cultivated in agar-agar and then heated up. This process emulated techniques used in Brazil and Japan since the beginning of the century, which some Pasteur researchers acknowledged. In addition, to support the idea that the antiplague vaccine manufactured in Paris could reach its objectives, Simond and also Édouard Dujardin-Beaumetz, in charge of producing the antiplague vaccine in the Parisian laboratory, sometimes referenced the results obtained in Japan and in Brazil with the vaccine cultivated in agar-agar. According to them, the work of Shiga and Cruz, which they knew by reading and by personal interaction, proved the efficacy of the antiplague vaccine that would now be delivered by the French laboratory.[73] The new antiplague vaccine manufactured in Paris was first used in different regions of the French empire during the 1910s, mostly Indochina and Senegal, before being injected on at least 1000 persons in Paris during the small plague outbreak known as *la peste des chiffonniers*, in 1920. After this epidemic, the antiplague vaccine cultivated in agar-agar became one of the main tools to fight this disease in the French empire, settling, at least from the point of view of the Pasteur Institute, a controversy that started in Bombay in 1897.[74]

Conclusion

People living in the two decades preceding World War I experienced an acceleration in globalization. The world, for dwellers of the largest cities of that period, seemed to be getting smaller, a sensation produced by the intensification of European imperialism in Africa and Asia, by the development of new forms of industrialization and communication devices, and by the development of international commerce. In part because of these correlated processes, the period was also marked by a new era of scientific activity and the emergence of new scientific disciplines, such as microbiology. In addition, the globalization paved the way for the (re)emergence and spread of old and new pathogens.

The imbricated process brought by the plague pandemic and the global development of microbiology led to the establishment of new, or the strengthening of older scientific connections between researchers and institutions based in different parts of the world, such as Brazil, France, India, and Japan. These interactions prompted a dynamic and global process of elaboration and reconfiguration of objects and practices in the field of microbiology. Knowledge about treatment and prevention of the plague, represented by sera and vaccines, the heralds of microbiology, was constructed mostly outside Europe and non-European actors played an important role there. This new knowledge was produced in many forms, ranging from the manufacturing of original antiplague sera and vaccines, through the application of innovative public health measures like mixed vaccination, to the amassing and circulation of annual sanitary statistics on serotherapy and immunization. The results of these different dynamics developed mainly outside Europe were often capable of reshaping scientific discourse and sanitary practices in the Old World. This is evidenced by the changes in the policies developed by the Pasteur Institute in the case of antiplague serotherapy and immunization thanks to their active and, at times, informal interactions with other centers researching the same subjects. In summary, the first years of the plague pandemic led to the global emergence of new actors in the field of microbiology, whose works reshaped the discipline. As a result, microbiology became, maybe for the first time, a truly global science.

In following the material circulation of people, texts, sera, and vaccines that evidenced this global emergence of microbiology, we have likewise described the shaping of a particular space of circulation of knowledge that connected actors based in France, Japan, India, and Brazil. It is true that this space did not *directly* connect all the players, and Brazil and Japan appeared here as two opposite sides of a picture, or of a map, connected with India and France, but not directly with each other. Nevertheless, it is worth noting that both the pandemic and microbiology produced a *conjuncture* that affected both Japan and Brazil and allowed them to act on a global scale. Indeed, both countries experienced similar questions, such as the doubts about the efficacy of serotherapy, the opportunity to send their sera to other countries and the necessity to explore techniques of immunization and shape them according to their local needs. Thus, we hope to have pointed to possible ways to write a global history of the emergence of microbiology, not as the result of scientific diffusion from Europe or the belated emergence of independent scientific traditions in the "Rest," but rather as a global scientific enterprise that can be understood by stressing global comparisons, connections, and circulation.

Notes

1   Claire Salomon-Bayet, "Penser la révolution pastorienne," in Claire Salomon-Bayet, ed., *Pasteur et la révolution pastorienne* (Paris: Payot, 1986), 15–53; Bruno Latour, *Pasteur: guerre et paix des microbes* (Paris: La Découverte, 2001); Ilana Löwy, "Cultures de bactériologie en France, 1880–1900: la paillasse et la politique," *Gesnerus* 67, no. 2 (2010): 188–216.

2   John Andrew Mendelsohn, *Cultures of Bacteriology: Formation and Transformation of a Science in France and Germany, 1870–1914* (PhD dissertation: Princeton University, 1996); Christoph Gradmann, *Laboratory Disease: Robert Koch's Medical Bacteriology* (Baltimore, MD: Johns Hopkins University Press, 2009); Silvia Berger, *Bakterien in Krieg und Frieden: Eine Geschichte der medizinischen Bakteriologie in Deutschland 1890–1933* (Göttingen: Wallstein Verlag, 2009).

3   Gabriel Gachelin, "The Designing of Anti-Diphtheria Serotherapy at the Institut Pasteur (1888–1900): The Role of a Supranational Network of Microbiologists," *Dynamis* 27 (2007): 45–62; Volker Hess, "The Administrative Stabilization of Vaccines: Regulating the Diphtheria Antitoxin in France and Germany, 1894–1900," *Science in Context* 21 (2008): 201–227; Christoph Gradmann and Jonathan Simon, *Evaluating and Standardizing Therapeutic Agents, 1890–1950* (Basingstoke: Palgrave Macmillan, 2010).

4   For the case of the French colonies, see Aro Velmet, *Pasteur's Empire: Bacteriology and Politics in France, its Colonies and the World* (Oxford: Oxford University Press, 2020). For India, see Pratik Chakrabarti, *Bacteriology in British India: Laboratory Medicine and the Tropics* (Rochester: University of Rochester, 2012). For Brazil, see Jaime Benchimol, *Dos micróbios aos mosquitos: febre amarela e a revolução pasteuriana no*

Brasil (Rio de Janeiro: Editora Fiocruz; Editora UFRJ, 1999) and Ilana Löwy, *Virus, moustiques et modernité: la fièvre jaune au Brésil, entre science et politique* (Paris: Éd. des Archives Contemporaines, 2001).

5 Kapil Raj, Beyond Postcolonialism ... and Postpositivism: Circulation and the Global History of Science," *Isis* 104, no. 2 (juin 2013): 337–347; Kapil Raj, Networks of Knowledge, or Spaces of Circulation? The Birth of British Cartography in Colonial South Asia in the Late Eighteenth Century," *Global Intellectual History* 2, no. 1 (2 janvier 2017): 49–66.

6 About the antidiphtheric serum and this new hope, see Esteban Rodríguez-Ocaña, "The Social Production of Novelty: The Diphtheria Serotherapy, 'herald of the New Medicine'," *Dynamis* 27 (2007): 21–44; Anne I. Hardy, "From Diphtheria to Tetanus: The Development of Evaluation Methods for Sera in Imperial Germany," in *Evaluating and Standardizing Therapeutic Agents*, 52–70.

7 Sylvia Chiffoleau, *Genèse de la santé publique internationale: de la peste d'Orient à l'OMS* (Rennes: Presses Universitaires de Rennes, 2012).

8 Myron J. Echenberg, *Plague ports: the global urban impact of bubonic plague, 1894–1901* (New York: New York University Press, 2007); Sylvia Chiffoleau, *Genèse de la santé publique internationale: de la peste d'Orient à l'OMS* (Rennes: Presses Universitaires de Rennes, 2012).

9 Alexandre Yersin, "La peste bubonique à Hong-Kong," *Annales de l'Institut Pasteur* VIII (1894): 662–667; Alexandre Yersin, Albert Calmette, and Amédée Borrel, "La peste bubonique. Deuxième note," *Annales de l'Institut Pasteur* IX, no. 7 (1895), 589–592; Alexandre Yersin, "Rapport sur la mission à Canton et Amoy," 1896, Archives de l'Institut Pasteur, FR AIP 6 YER.

10 "Report by Surgeon-Major Lyons, I.M.S., President of the Plague Research Committee," 1897, IP/13/PC.2, National Library of Scotland.

11 Alexandre Yersin, "Rapport sur la peste aux Indes," *Archives de Médecine Navale* 68 (1897): 366–372; Paul-Louis Simond, "Rapport sur les cas de peste traités dans l'Inde Anglaise par le moyen du sérum antipesteux préparé à l'Institut Pasteur de Paris par M. le Dr Roux," *Annales d'hygiène et de médecine coloniales* 1 (1898): 349–361.

12 The Italian serum was manufactured in a different manner than the French serum: instead of injecting the plague bacilli directly in the horses – the Pasteur Institut method – Lustig and Galeotti applied in these animals the nucleoprotein of the plague bacilli, a safer method according to them. See Alessandro Lustig and Gino Galeotti, "On the Vaccination of Animals against the Plague Bacillus, and on the Serum Obtained Therefrom," *British Medical Journal* 1 (1897): 1027–1028; Gino Galeotti and Giovanni Polverini, *Sui primi 175 casi di peste bubbonica tratati nel 1898 in Bombay col siero preparato nel Laboratório di Patologia Generale di Firenze* (Florence: Carnesecchi e Figli, 1898).

13 Gino Galeotti, "Il Laboratorio municipale di Bombay per la preparazione del siero contro la peste bubbonica," *Rivista d'Igiene e Sanità Publica*, Suplemento al n. 7 (1899): 1–8; "Letter from the Acting Municipal Commissioner for the City of Bombay to the Municipal Secretary, N° 10,895," July 10, 1902, IOR/P/6477 page 359, British Library, India Office Records.

14 Albert Calmette and Alexandre Salimbeni, "La peste bubonique: étude de l'épidémie d'Oporto en 1899 (sérothérapie)," *Annales de l'Institut Pasteur* XIII, no. 12 (1899), 866–936.

15 Alessandro Lustig and Gino Galeotti, "The Prophylactic and Curative Treatment of Plague," *The British Medical Journal* 1 (1901): 206–208.

16 "Shiga igaku hakase pesuto chôsa ni kansuru enzetsu," 13 August 1898, no. 287-10, Archives of the Institute of Taiwan History, Academia Sinica, Taiwan Sôtokufu Official Records.

17 Shibasaburô Kitasato, "Pesutobyô chôsa ihô," *Saikingaku zasshi* no. 50 (1900): 65–67.

18 "Yersin hakase no kessei kuru," *Dainihon shiritsu eiseikai zasshi* no. 199 (1899), 978.

19 "France Indochina sôtoku no kôi," *Dainihon shiritsu eiseikai zasshi* no. 200 (1900): 56–57.

20 Masahiro Iwai, "Osaka ni okeru pesto no saikingakuteki kenkyû," *Saikingaku zasshi* no. 53 (1900): 231–240.

21 Sahachirô Hata, "Pesuto saikingaku bumon," *Saikingaku zasshi* no. 54 (1900): 43–69.

22 Ki'ichi Tsukiyama, "Tainan-shi ni okeru pestuo kessei ryôhô seiseki," *Saikingaku zasshi* no. 71 (1901): 736–768.

23 Sahachirô Hata, "Kessei ryôhô wo okonaeru pesuto kanja no ondohyô demonstration," *Saikingaku zasshi* no. 64 (1901): 201–202.

24 Shôta Matsumoto, "Taishô san nen Tokyo ni okeru pesuto no rinshô tekin narabini saikingaku teki kansatsu," *Saikingaku zasshi* no. 237 (1915): 467–482.

25 Gorôsaku Shibayama, "Pesuto chiryô kessei ni kansuru hôkoku," *Saikingaku zasshi* no. 147 (1908): 122–123.

26 Gorôsaku Shibayama, "Manshû ni okeru hai pesuto ryûkô," *Saikingaku zasshi* no. 189 (1911): 471–506.

27 Richard Strong, *Report of the International Plague Conference held at Mukden, April, 1911* (Manila: Bureau of Printing, 1912).

28 Kaoru Katayama, "Dairen ryô byôin ni okeru hai pesuto kanja chiryô seiseki ni tsuite," *Saikingaku zasshi* no. 191 (1911): 672–681.

29 "The results of the international plague conference of Mukden. Official Statement by the Waiwu Pu," May 4, 1911, no. 5430-22, Archives of the Institute of Taiwan History, Academia Sinica, Taiwan Sôtokufu Official Records.

30 Benchimol, *Dos micróbios aos mosquitos*.

31 Matheus Alves Duarte da Silva, *Quand la peste connectait le monde: production et circulation de savoirs microbiologiques entre Brésil, Inde et France (1894–1922)* (PhD dissertation, École des Hautes Études en Sciences Sociales, 2020).

32 Camillo Terni, "Vacinação e soroterapia da peste bubônica," *Brazil-Médico* XIV, no. 17 (1900): 143–152; Victor Godinho, "Contribuição para o estudo clínico da peste," in Quarto Congresso Brasileiro de Medicina e Cirurgia. *Annaes.* (Rio de Janeiro: Imprensa Nacional, 1902), 128–129.

33 Camillo Terni, "The Plague in Rio de Janeiro," IOR/L/E/7/440, British Library, India Office Records.

34 Gonçalo Moniz de Aragão, *Relatório sobre a preparação da vaccina e do sôro antipestilentos* (Salvador: Typographia e Encadernação do Diário da Bahia, 1902).

35 Letter from Vital Brazil to the Under-secreatry of State for India, August 1902, Correspondência Científica, Carta n° 205, Arquivos do Instituto Butantan; Henrique Figueiredo de Vasconcellos, "O soro anti-pestozo," *Memórias do Instituto Oswaldo Cruz* 1 (1909): 14–27.

36 "Translation from Portuguese in Notes on the Plague Serum, By Dr. Vital Brazil, Director of the São Paulo Institute," Londres, May 19, 1902, IOR/L/E/7/464, British Library, India Office Records; Antonino Ferrari, "A sorotherapia na peste: Injeções endovenosas em 69 doentes - Mortalidade 7,2%," *Revista Médica de São Paulo* X, no. 23 (1907): 491–496; Henrique Figueiredo de Vasconcellos, "Prophylaxie de la peste à Rio de Janeiro," *Annales de l'Institut Pasteur* XXII, no. 10 (1908): 819–831.

37 Abdon Petit-Carneiro, "Ação curativa do serum anti-pestifero de Butantan," *Revista Médica de São Paulo* V, no. 4 (1902): 65–69.

38 Antonio Cardoso Fontes, *Vacinação e soroterapia antipestosas* (Rio de Janeiro: Leon de Rennes, 1903), 90–91. Oswaldo Cruz, "Relatório do Diretor Geral de Saúde Pública," in *Ministério da Justiça e Negócios Interiores. Ministro J. J. Seabra. Relatório dos anos 1903 e 1904 apresentado ao Presidente da República dos Estados Unidos do Brazil* (Rio de Janeiro: Imprensa Nacional, 1904).

39 "Plague serum from the serum-therapeutic Institut, São Paulo, Brazil, Letter from Dr W. Strain to the India Office," May 19, 1902, IOR/L/E/7/464, British Library, India Office

Records; Vital Brazil, "Letter from to the Under-Secretary of State for India," June 16, 1903, Correspondência Científica, Carta n° 304, Arquivos do Instituto Butantan.

40 Waldemar Haffkine, *Serum-Therapy of Plague in India; Reports by Mr. W. M. Haffkine, c.i.e., and Various Officers of the Plague Research Laboratory, Bombay. Edited with an introduction by Lieut.-col. W. B. Bannerman* (Calcutta: Office of the Superintendent of Government Printing, 1905); Nasarwanji Choksy, "On Recent Progress in the Serum-therapy of Plague," *British Medical Journal* 1 (1908): 1282–1284.

41 Camillo Terni, *Studi sulla peste. Parte 2: Cura razionale della peste* (Milano: Unione tipografico-editrice torinese, 1906).

42 Dorival de Camargo Penteado, "Tratamento da Peste," *Revista Médica de São Paulo* XI, no. 5 (1908): 89–99.

43 Abdon Petit-Carneiro, "Ação curativa;" Antonio Cardoso Fontes, *Vacinação e soroterapia*; Oswaldo Cruz, "Relatório;" Henrique Figueiredo de Vasconcellos, "Prophylaxie."

44 Émile Roux, "Manuscrit du Cours de Microbic Technique, 41e leçon, Peste Bubonique (2e leçon)," 1908–1909, Cote: B1-3, Archives de l'Institut Pasteur; Edouard Dujardin-Beaumetz, "Sérothérapie et vaccination de la peste bubonique," in *Bibliothèque de thérapeutique* (Paris: B. Baillière et fils, 1909).

45 Waldemar Haffkine, "Remarks on the Plague Prophylactic Fluid," *British Medical Journal* 1 (1902): 1461–1462.

46 William Bannerman, *Report of the Plague Research Laboratory for the Official Year Ending 31st March 1905* (Bombay: Government Central Press, 1906); E. C. Hodgson, *Report of the Bombay Bacteriological Laboratory for the year 1920* (Bombay: Government Central Press, 1922).

47 Waldemar Haffkine, *A conversazione on the inoculation against plague held by Mons. W. M. Haffkine in the Cantonment Magistrate Office, in Puna* (Poona: s.n., 1898); William Bannerman, "The Plague Research Laboratory of the Government of India, Parel, Bombay," *Proceedings of Royal Society of Edinburgh* XXIV (1902): 113–144.

48 Yersin, "Rapport sur la peste aux Indes," "Lettre de Paul-Louis Simond au Surgeon General," 1898, Cote: SIM.5, Archives de l'Institut Pasteur; Paul-Louis Simond, "La propagation de la peste," *Annales de l'Institut Pasteur* XII, no. 10 (1898): 625–687.

49 Alexandre Yersin, "L'épidémie de peste à Nha-Trang de à octobre 1898," *Annales d'hygiène et de médecine coloniales* 2 (1899): 373–385; Adrien Proust, "Sur la police sanitaire maritime et le séjour du Sénegal au lazaret du Frioul," *Bulletin de l'Académie Nationale de Médecine* 36 (1901): 488–503.

50 Calmette and Salimbeni, "La peste bubonique."

51 Terni, *Studi sulla peste.*

52 The discussion between Cruz and Terni can be found at Quarto Congresso Brasileiro de Medicina e Cirurgia, *Annaes*, 36–38.

53 Georg Gaffky et al., *Bericht über die Thätigkeit der zur Erforschung der Pest im Jahre 1897 nach Indien entsandten Kommission* (Berlin: Julius Springer, 1899).

54 Oswaldo Cruz, "Trabalho com a vacina antipestosa," BR.RJ.COC.OC.PI.TP.53, Arquivos da Casa de Oswaldo Cruz; Oswaldo Cruz, "A vacinação antipestosa," *Brazil-Médico* XV, no. 45 (1901): 443–447; Oswaldo Cruz, "A vacinação antipestosa (continuação)," *Brazil-Médico* XV, no. 47 (1901): 463–467; Oswaldo Cruz, "A vacinação antipestosa (conclusão)," *Brazil-Médico* XV, no. 48 (1901): 473–477.

55 Jaime Benchimol, "Reforma urbana e revolta da vacina na cidade do Rio de Janeiro," in Jorge Delgado Ferreira and Lucilia de Almeida Neves, eds., *O Brasil republicano: o tempo do liberalismo excludente – da proclamação da república à revolução de 1930* (Rio de Janeiro: Civilização Brasileira, 2003), 231–286.

56 Oswaldo Cruz, *Dos accidentes em serotherapia* (Rio de Janeiro: Typ. Besnard Frères, 1902); Oswaldo Cruz, "Peste," in *Formulário Prático do Brazil-Médico*, 1906; Henrique Figueiredo de Vasconcellos, "Prophylaxie."

57 Henrique Figueiredo de Vasconcellos, "Prophylaxie."

58 Brasil, *Ministério da Justiça e Negócios Interiores. Ministro Rivadavia da Cunha Corrêa. Relatório dos anos de 1912 e 1913 apresentado ao Presidente da República dos Estados Unidos do Brazil. Diretoria do Interior* (Rio de Janeiro: Imprensa Nacional, 1913), 21.

59 Iwai, "Osaka ni okeru pesto no saikingakuteki kenkyû."

60 Shibasaburô Kitasato et al., "Kobe-shi Osaka-shi pesutobyô chôsa hôkoku : 2," *Saikingaku zasshi* no. 57 (1900): 453–488.

61 Iwai, "Osaka ni okeru pesto no saikingakuteki kenkyû," 238

62 Kiyoshi Shiga, "Sekiribyô kenkyû hôkoku 1," *Saikingaku zasshi* no. 25 (1897): 787–810.

63 Kiyoshi Shiga, "Sekiri yobôchûsha-hô yohô," *Saikingaku zasshi* no. 46 (1899): 493–495.

64 Shiga did not mention animal experiments in his article on dysentery vaccine. The safety of the vaccine was demonstrated through his body itself: he tested his first vaccine on himself. See Takeshi Odaka, *Densenbyô kenkyûjo : kindai igaku kaitaku no michinori* (Tokyo: Gakkai shuppan sentâ, 1992).

65 Kiyoshi Shiga, "Pesuto yobôsesshuhô ihô," *Saikingaku zasshi* no. 50 (1900): 27–28.

66 Gorôsaku Shibayama and Sahachirô Hata, "Pesuto yobô sesshu ni tsuite," *Saikingaku zasshi* no. 65 (1901): 231–265.

67 "Anping zeikan-in pesto ni kakari Tainan iin ni oite hoka no kenzensha ni yobôeki chûsha jôkyô," February 1, 1900, no. 4643-1, Archives of the Institute of Taiwan History, Academia Sinica, Taiwan Sôtokufu Official Records.

68 Gorôsaku Shibayama and Sahachirô Hata, "Pesuto yobô sesshu ni tsuite," 252.

69 The document is dated as written in November 1901. But given the explained context and considering that this discourse is found in an article that appeared on *Saikingaku Zasshi* at the beginning of 1901, we consider that the letter was written in November 1900. "A letter from Ki'ichi Tsukiyama to Naoshi Katô," November 28, 1901, no. 4643-1, Archives of the Institute of Taiwan History, Academia Sinica, Taiwan Sôtokufu Official Records.

70 Ki'ichi Tsukiyama, "Tainan-ken ni okeru pesuto yobô sesshu seiseki dai ni hô," *Saikingaku zasshi* no. 71 (1901): 724–736.

71 Alexandre Besredka, "De l'immunisation active contre la peste, le choléra et l'infection typhique," *Annales de l'Institut Pasteur* XVI, no. 12 (1902): 918–930.

72 Alexandre Besredka, "De l'anaphylaxie sérique expérimentale," *Bulletin de l'Institut Pasteur* no. 19 (1908): 841–852; Ilana Löwy, "On Guinea Pigs, Dogs and Men: Anaphylaxis and the Study of Biological Individuality, 1902–1939," *Studies in History and Philosophy of Science Part C: Studies in History and Philosophy of Biological and Biomedical Sciences* 34 (2003): 399–423.

73 Édouard Dujardin-Beaumetz, "Sérothérapie et vaccination," 344–345; Paul-Louis Simond, "Peste," in *Traité d'hygiène* (Paris: J.B. Baillière et Fils, 1911), 526.

74 Édouard Dujardin-Beaumetz, "Technique de la vaccination antipesteuse," *Le Journal Médical Français* 10, no. 1 (1921): 86; Edouard Joltrain, *La peste : étiologie, formes cliniques, prophylaxie & traitement* (Paris: A. Maloine et fils, 1921).

Index

Note: Page numbers in italics refer to figures and those followed by "n" refer to notes.

For Product Safety Concerns and Information please contact our EU
representative GPSR@taylorandfrancis.com
Taylor & Francis Verlag GmbH, Kaufingerstraße 24, 80331 München, Germany